A Definitive Guide to Apache ShardingSphere

Transform any DBMS into a distributed database
with sharding, scaling, encryption features, and more

Trista Pan

Zhang Liang

Yacine Si Tayeb, PhD

BIRMINGHAM—MUMBAI

A Definitive Guide to Apache ShardingSphere

Copyright © 2022 Packt Publishing

Publishing Product Manager: Devika Battike
Senior Editors: Roshan Kumar and Tazeen Shaikh
Content Development Editor: Shreya Moharir
Technical Editor: Rahul Limbachiya
Copy Editor: Safis Editing
Project Coordinator: Farheen Fathima
Proofreader: Safis Editing
Indexer: Rekha Nair
Production Designer: Shyam Sundar Korumilli
Marketing Coordinator: Nivedita Singh

First published: July 2022
Production reference: 1300622

Published by Packt Publishing Ltd.
Livery Place
35 Livery Street
Birmingham
B3 2PB, UK.

ISBN 978-1-80323-942-2
www.packt.com

Contributors

About the authors

Trista Pan is the co-founder and CTO of SphereEx, an Apache Member and Incubator Mentor, Apache ShardingSphere PMC, AWS Data Hero, China Mulan open source community mentor, and Tencent Cloud TVP. Trista used to be responsible for the design and development of the intelligent database platform of JD Digital Science and Technology. She now focuses on the distributed database and middleware ecosystem, and the open source community. She was the recipient of the *2020 China Open-Source Pioneer*, *2021 OSCAR 2021 Top Open Source Pioneer*, and *2021 CSDN IT Leading Personality* awards. Her paper, *Apache ShardingSphere: A Holistic and Pluggable Platform for Data Sharding*, was published on ICDE in 2022.

Zhang Liang is the founder and CEO of SphereEx, an Apache Member, the founder of Apache ShardingSphere ElasticJob, the PMC Chair, Tencent Cloud TVP, and Microsoft MVP. Zhang is an open source enthusiast and thought leader in Java-based distributed architectures. Currently, he focuses on turning Apache ShardingSphere into an industry-leading distributed database solution. His 2019 book, *Future Architecture: From Service to Cloud Native*, was well received by both critics and the community. His 2022 paper, *Apache ShardingSphere: A Holistic and Pluggable Platform for Data Sharding*, was published on ICDE. Zhang was awarded titles in the *Top Ten Distributed Database Pioneers of 2021* by CSDN, and *the 33 China Open Source Pioneers in 2021* by SegmentFault.

Yacine Si Tayeb, PhD, is the Head of International Operations at SphereEx and one of the core contributors and community builders at Apache ShardingSphere. Passionate about technology and innovation, Yacine moved to Beijing to pursue his PhD in enterprise management and was in awe of the local startup and tech scene. His career path and research have so far been shaped by opportunities at the intersection of technology and business. He is a published scholar, and his passion for technology led him to research the impact of corporate governance and financial performance on corporate innovation outcomes, and to take a keen interest in the development of the Apache ShardingSphere big data ecosystem and open source community building.

About the reviewers

Longtao Jiang is an Apache ShardingSphere committer. He has been active in the community for a long time and has contributed many useful functions to ShardingSphere. Before becoming a committer, he was also a skilled user of ShardingSphere, applying ShardingSphere to multiple financial-level data sharding scenarios. Now, Longtao Jiang is mainly responsible for DistSQL-related innovations and practices in the ShardingSphere community.

Zhengqiang Duan is a committer of the Apache ShardingSphere community and is also a senior middleware engineer at SphereEx. He has been in contact with the Apache ShardingSphere project since 2018 and has led data sharding projects with massive amounts of data within the company and has rich practical experience. He loves open source very much and is willing to share and communicate. He is currently focusing on the development of Apache ShardingSphere kernel modules and strives to provide more powerful and easy-to-use features for the Apache ShardingSphere community.

Nianjun Sun has 15 years of coding experience as a Java developer and is interested in cloud-native and distributed database-related technology. He used to be the architect of Bizseer and was responsible for the design and development of their AIOps platform. He currently works in the core team of Apache ShardingSphere, which founded and built distributed data infrastructures while delivering a SaaS experience through the cloud. He published the paper, *Apache ShardingSphere: A Holistic and Pluggable Platform for Data Sharding*, on ICDE in May 2022.

Hongsheng Zhong is an Apache ShardingSphere committer and is passionate about open source and database ecosystems. Currently, he works for SphereEx as a senior Java engineer, focusing on the development of the Apache ShardingSphere database middleware ecosystem and open source community building. Previously, he worked on R&D of cloud database products at JD Technology, and has experience of multiple replicas on Raft.

Table of Contents

Section 2: Apache ShardingSphere Architecture, Installation, and Configuration

3

Key Features and Use Cases – Your Distributed Database Essentials

4

Key Features and Use Cases – Focusing on Performance and Security

5

Exploring ShardingSphere Adaptors

6

ShardingSphere-Proxy Installation and Startup

7

ShardingSphere-JDBC Installation and Start-Up

Section 3: Apache ShardingSphere Real-World Examples, Performance, and Scenario Tests

8

Apache ShardingSphere Advanced Usage – Database Plus and Plugin Platform

9

Baseline and Performance Test System Introduction

10

Testing Frequently Encountered Application Scenarios

11

Exploring the Best Use Cases for ShardingSphere

12
Applying Theory to Practical Real-World Examples

Appendix and the Evolution of the Apache ShardingSphere Open Source Community

Index
Other Books You May Enjoy

Preface

Apache ShardingSphere is a new open source ecosystem for distributed data infrastructures based on pluggability and cloud-native principles.

This book begins with a quick overview of the main challenges faced by DBMSs today in production environments, followed by a brief introduction to the ShardingSphere software's kernel concept. Thereafter, through real-world examples, including distributed database solutions, elastic scaling, DistSQL, synthetic monitoring, SQL authorization and user authentication, and database gateway, you will gain a full understanding of ShardingSphere's architectural components and how they are configured and can be plugged into your existing infrastructure to manage your data and applications. Moving ahead, you will get well versed with ShardingSphere-JDBC and ShardingSphere-Proxy, the ecosystem's clients, and how they can work either concurrently or independently according to your needs. Then, you will learn how to customize the plugin platform to define your personalized user strategies and manage multiple configurations seamlessly. Lastly, you will get up and running with functional and performance tests for all scenarios.

By the end of this book, you will be able to build and deploy your own customized version of ShardingSphere, addressing the key pain points encountered in your data management infrastructure.

Who this book is for

This book is for database administrators (DBAs) working with distributed database solutions who are looking to explore the capabilities of Apache ShardingSphere. DBAs looking for more capable, flexible, and cost-effective alternatives to the solutions they're currently utilizing will also find this book helpful. To get started with this book, a basic understanding of, or even an interest in, databases, relational databases, SQL, cloud computing, and data management in general is needed.

What this book covers

Chapter 1, The Evolution of DBMSs, DBAs, and the Role of Apache ShardingSphere, introduces the main challenges faced by DBMSs today in production environments, the evolving role of the DBAs, and the opportunities and future directions for DBMSs. This chapter lays the foundation for the rest of this book by including a brief introduction to the Apache ShardingSphere ecosystem, the context of the project development, and the need filled by the software solution.

Chapter 2, Architectural Overview of Apache ShardingSphere, provides a professionally focused description of the software's architecture. The Database Plus driving development concept is introduced, together with the deployment architecture and the plugin platform.

Chapter 3, Key Features and Use Cases – Your Distributed Database Essentials, gives an overview of the potential use cases of ShardingSphere in professional and enterprise environments divided by industry (fintech, media, e-commerce, etc.) and introduces the solution's features that are necessary for a distributed database.

Chapter 4, Key Features and Use Cases – Focusing on Performance and Security, expands your knowledge on the potential use cases of ShardingSphere in a professional/enterprise environment by focusing on the ecosystem's features that'll allow you to monitor and improve performance and enhance security.

Chapter 5, Exploring ShardingSphere Adaptors, includes a description of the ecosystem's main clients, their differences, and how they can work either concurrently or independently, depending on your needs.

Chapter 6, ShardingSphere-Proxy Installation and Startup, introduces ShardingSphere-Proxy, how to use it directly as MySQL and PostgreSQL servers, and how to apply it to any kind of terminal.

Chapter 7, ShardingSphere-JDBC Installation and Startup, explains the client end, how it connects to databases, its third-party database connection pool, and how to successfully install it.

Chapter 8, Apache ShardingSphere Advanced Usage – Database Plus and Plugin Platform, illustrates how to customize the plugin platform on an ad hoc basis to get the most out of your system and introduces cloud-native principles.

Chapter 9, Baseline and Performance Test System Introduction, introduces the built-in baseline and performance testing system and how to use it – from preparing your test to report analysis.

Chapter 10, Testing Frequently Encountered Application Scenarios, offers tests for frequently encountered application scenarios, including distributed databases, read/write splitting, and shadow databases.

Chapter 11, Exploring the Best Use Cases for ShardingSphere, showcases the best use cases for each scenario, as well as a series of real-world examples such as distributed database solutions, database security, synthetic monitoring, and database gateway.

Chapter 12, Applying Theory to Practical Real-World Examples, builds on the knowledge cumulated in *Chapter 11, Exploring the Best Use Cases for ShardingSphere*, and provides methodologies to turn theory into practice.

Appendix and the Evolution of the Apache ShardingSphere Open Source Community, offers a guide to leverage the ecosystem's documentation, the example projects in the GitHub repository, more information on ShardingSphere's source code and license, and how to join the project's open source community.

To get the most out of this book

For this book, to be able to get the most out of your system and Apache ShardingSphere, you will need a few simple tools.

We have compiled a list of the software you may need, depending on the features or ShardingSphere clients you are interested in.

In terms of the operating system, you may use any of the mainstream choices available, Windows, macOS, or Linux. All code examples have been tested in all three operating systems and should work with future versions too.

Software/hardware covered in the book	Operating system requirements
ShardingSphere-JDBC 5.0.0	Windows, macOS, or Linux
ShardingSphere-Proxy 5.0.0	
Visual Studio Code (Text Editor)	
MySQL 5.7 +	
MySQL Workbench (MySQL GUI client)	
PostgreSQL 12 +	
PgAdmin4 (PostgreSQL GUI client)	
ZooKeeper 3.6 +	
PrettyZoo (ZooKeeper GUI)	
JRE/JDK 8 +	
Docker	
Sysbench 1.0.20	

If you were to encounter any difficulties in the installation, and you cannot find the fix covered in this book for reasons such as a particular setup you may be using, you can reach out to the community in the *Issues* or *Discussions* sections of Apache ShardingSphere's GitHub repository.

If you are using the digital version of this book, we advise you to type the code yourself or access the code from the book's GitHub repository (a link is available in the next section). Doing so will help you avoid any potential errors related to the copying and pasting of code.

ShardingSphere uses the Apache Software Foundation's Apache License 2.0. You can find the complete details here: `https://www.apache.org/licenses/LICENSE-2.0`.

Download the example code files

You can download the example code files for this book from GitHub at `https://github.com/PacktPublishing/A-Definitive-Guide-to-Apache-ShardingSphere`. If there's an update to the code, it will be updated in the GitHub repository.

We also have other code bundles from our rich catalog of books and videos available at `https://github.com/PacktPublishing/`. Check them out!

You can connect with the community for more events, and user cases, and if you want to try, become an open source developer.

Download the color images

We also provide a PDF file that has color images of the screenshots and diagrams used in this book. You can download it here: `https://packt.link/VUBd8`.

Conventions used

There are a number of text conventions used throughout this book.

`Code in text`: Indicates code words in text, database table names, folder names, filenames, file extensions, pathnames, dummy URLs, user input, and Twitter handles. Here is an example: "Mount the downloaded `WebStorm-10*.dmg` disk image file as another disk in your system."

A block of code is set as follows:

```
html, body, #map {
  height: 100%;
  margin: 0;
  padding: 0
}
```

When we wish to draw your attention to a particular part of a code block, the relevant lines or items are set in bold:

```
[default]
exten => s,1,Dial(Zap/1|30)
exten => s,2,Voicemail(u100)
exten => s,102,Voicemail(b100)
exten => i,1,Voicemail(s0)
```

Any command-line input or output is written as follows:

```
$ mkdir css
$ cd css
```

Bold: Indicates a new term, an important word, or words that you see onscreen. For instance, words in menus or dialog boxes appear in **bold**. Here is an example: "Select **System info** from the **Administration** panel."

> **Tips or important notes**
> Appear like this.

Get in touch

Feedback from our readers is always welcome.

General feedback: If you have questions about any aspect of this book, email us at customercare@packtpub.com and mention the book title in the subject of your message.

Errata: Although we have taken every care to ensure the accuracy of our content, mistakes do happen. If you have found a mistake in this book, we would be grateful if you would report this to us. Please visit www.packtpub.com/support/errata and fill in the form.

Piracy: If you come across any illegal copies of our works in any form on the internet, we would be grateful if you would provide us with the location address or website name. Please contact us at copyright@packt.com with a link to the material.

If you are interested in becoming an author: If there is a topic that you have expertise in and you are interested in either writing or contributing to a book, please visit authors.packtpub.com.

Share Your Thoughts

Once you've read *A Definitive Guide to Apache ShardingSphere*, we'd love to hear your thoughts! Scan the QR code below to go straight to the Amazon review page for this book and share your feedback.

https://packt.link/r/1-803-23942-5

Your review is important to us and the tech community and will help us make sure we're delivering excellent quality content.

Section 1: Introducing Apache ShardingSphere

In this part, you will gain an overview of Apache ShardingSphere, its architecture, concepts, and clients. You will get up to speed on the latest challenges affecting databases today and future developments, and be able to conceptualize ShardingSphere's position in the current database landscape.

This section comprises the following chapters:

- *Chapter 1, The Evolution of DBMSs, DBAs, and the Role of Apache ShardingSphere*
- *Chapter 2, Architectural Overview of Apache ShardingSphere*

1

The Evolution of DBMSs, DBAs, and the Role of Apache ShardingSphere

Today, data is recognized as the most valuable property available. As the so-called *warehouses* for this most valuable property, databases were not always given the enviable amount of attention they have been getting as of late. The hyper-growth of the internet, as well as its related and non-related industries (think traditional sectors affected by the positive externalities of increased connectivity, such as transportation and retail), the emergence of cloud-native, the development of the database industry, and distributed technology, have brought up new requirements and renewed pressure on businesses and their infrastructure.

Additionally, changes in societies at large, coupled with changes to people's lifestyles, have also raised new issues, concerns, and requirements for any modern company. Accordingly, companies must review their products, services, and architectures for their end users and consider upgrading and innovating from the frontend to the backend. Ultimately, they must consider the database and data as the most vital parts of this evolutionary process.

Simply put, data drives businesses. Stakeholders from C-suite executives, such as CIOs, to database managers are aware of the important role that data plays in transforming their businesses, satisfying users, and allowing them to maintain or create new competitive advantages.

Such recognition created a focus on three key areas all related to data – data collection, data storage, and data security – all of which will be discussed in detail in this book. The absence of databases from this list is by no means a lack of appreciation toward their integral role within organizations, but only the omission of an obvious fact.

Overlooking databases can create inefficiencies that can quickly snowball and become seriously threatening problems, such as a poor database experience for employees and customers, cost overruns, and poor workload optimization. At the same time, enterprises also need capable experts to leverage their databases and manage and efficiently utilize the data. Hence, the **data**, **database**, and **database administrator** (**DBA**) form a system that allows enterprises to efficiently store, protect, and leverage their assets.

In this chapter, we will cover the following topics:

- The evolution of DBMSs
- The evolving role of the DBA
- The opportunities and future directions for DBMSs
- Understanding Apache ShardingSphere

By the end of this chapter, you will have developed a comprehensive understanding of the current challenges for DBMSs. For those of you that are already familiar with the ongoing evolution of the database industry, this chapter will serve either as a refresher of the most pressing challenges or as a reference that organizes these challenges for you into one place.

Understanding these challenges will be followed by an introduction to the Apache ShardingSphere ecosystem and its driving concepts. Finally, you will be able to answer how ShardingSphere can help you solve the most pressing DBMS challenges and support you well into the future evolution of the database industry.

The evolution of DBMSs

With the rapid adoption of the cloud, SaaS delivery models, and open source repositories that are driving innovation, the proliferation of data has exploded in the past 10 years. These large datasets have made it mandatory for organizations who want an optimal customer experience to deploy effective and reliable **database management systems (DBMSs)**. Nevertheless, this renewed focus for organizations on DBMSs and their requirements has not only created multiple opportunities for new technologies and new players in the industry but also numerous challenges. If you are reading this book, you are probably looking to upskill yourself and improve or expand your knowledge on how to effectively manage DBMSs.

Databases exist to store and access information. As a result, organizations now find it crucial to understand the latest techniques, technologies, and best practices to store and retrieve extensive data and the resulting traffic. The shift to cloud-based storage has also led to the expanded use of data clusters, and the related data science around data storing strategies. Data use for apps goes up and down throughout a typical day.

Reliable and scalable databases are required to help collect and process data by breaking large datasets into smaller ones. Such a need gave rise to concepts such as database sharding and partitioning, where both are used to scale extensive datasets into smaller ones while preserving performance and uptime. These concepts will be discussed in *Chapter 3, Key Features and Use Cases – Your Distributed Database Essentials*, in the *Understanding data sharding* section, and *Chapter 10, Testing Frequently Encountered Application Scenarios*.

Let's summarize what open source means according to *The Open Source Definition* (https://opensource.org/osd) – when we talk about open source, we refer to software that's released under a license where the copyright holder gives you and any other user the rights to use or change and distribute the software, even its source code, to anyone for any purpose deemed fit.

When it comes to databases, the role of open source is not only non-negligible, but it may come as a surprise to many. As of June 2021, over 50% of database management systems worldwide use an open source license (DB-Engines, Statista 2021). If we consider the recent developments of open source database software, we'll notice the proliferation of initiatives and communities dedicated to cloud-native database software.

Cloud-native databases have become increasingly important with the ushering-in of the cloud computing era. Its benefits include elasticity and the ability to meet on-demand application usage needs. Such a development creates the need for cloud migration capabilities and skills as businesses migrate workloads to different cloud platforms.

Currently, hybrid and multi-cloud environments are the norm, with nearly 75% of organizations reporting usage of a multi-cloud environment (`https://www.lunavi.com/blog/multi-cloud-survey-72-using-multiple-cloud-providers-but-56-have-no-multi-cloud-strategy`). The data that remains stored on-premises is, more often than not, composed of sensitive information that organizations are wary of migrating, or data that is connected to legacy applications or environments that make it too challenging to migrate.

This changed the concept of databases as we used to understand them, creating a new concept that includes data that is on-premises and in the cloud, with workloads running across various environments. The next big thing in terms of databases and infrastructure is the distributed cloud, which can be defined as an architecture where multiple clouds are used concurrently and managed centrally from a public cloud. It brings cloud-based services to organizations and blurs the lines between the cloud and on-premises systems.

The next section will introduce you to the challenges that are currently considered to be significant pain points in the industry. You may be familiar with some or all of them – if you are not, that is OK, and you will find that they are all explained in the next section.

These pain points will then be followed by equally important needs that currently haven't been met or are currently creating new opportunities in the industry.

Industry pain points

Because of the ever-expanding number of database types, engineers have to dedicate more of their time to learning SDKs and SQL dialects, and less time to developing. For an enterprise, technology selection is hard because of more complex tech stacks and the need to match their application frameworks, which can cause an oversized architecture.

The next few sections will introduce you to the most notable industry pain points, followed by new industry needs that are creating new opportunities for DBMSs.

Low-efficiency database management

Database administrators (**DBAs**) need to dedicate much of their time to surveying and using new databases to identify the differences in cooperation and monitoring methods, as well as to understand how to optimize performance.

The peripheral services and experience of a certain database are not universal or replicable. In production, the usage and maintenance cost of databases rises. The more database types a company deploys, the more investment will be required. If an enterprise adopts new databases suitable for new scenarios without a second thought, the investment is doomed to exponentially grow sooner or later.

New demands and increasingly frequent iteration

Different code is required to meet what could seem to be similar demands, with the only difference being the database type and the type of code that it supports. At the time of writing, while iteration frequency is already expected to rise sharply, developer response capability is reduced and inversely proportional to the number of database types. The exponential growth of common demands and database types slows down iteration significantly. The larger the number of databases, the slower the iteration pace and the lower the iteration performance level.

If, for example, the desired outcome is to encrypt all sensitive data at once, but doing so on a one-to-many database failed, the only possible solution is to modify the code on the business application side. Large firms frequently operate with dozens or even hundreds of systems, which poses great challenges for developers in encrypting all systems' data. Data encryption is only one of the many possible example challenges of this kind that developers may face, with other common demands such as permissions control, audit, and others all being frequently encountered in heterogeneous databases.

Lack of database inter-compatibility

We know for a fact that heterogeneous databases currently co-exist and will continue to do so for a long time, but without a common standard, we cannot collaboratively use databases. By common standard, we mean a universally accepted (or at least by a majority) technology reference such as the USB 2.0 or USB-C standard is for external hardware peripherals. If you are looking for a software example, look no further than SDKs that have been released to make apps for iOS or Android.

For databases, as you will learn throughout this book, we at the Apache ShardingSphere community are proponents of what we call **Database Plus** – which in simple terms means software that allows you to manage and improve any type of database, even to integrate different database types into the same system.

In terms of data computing, demands for a collaborative query engine and transaction management plans across heterogeneous databases are increasing. Nevertheless, at the moment, developers can only contribute to the development on the application side, making it difficult for their contribution to be developed into an infrastructure.

The new industry needs are creating new opportunities for DBMSs

The changing landscape within which enterprises operate is bound to affect their business decisions and operating procedures. This can be traced back to the expanding amounts of data and the internet argument mentioned in the *Industry pain points* section.

This section will give you an insight into what enterprises are looking to get from their database management systems across different industrial sectors. After that, we will look at the evolving role of a DBA, which some of you will be expected to step into.

Querying and storing enormous chunks of data

A large volume of data can crash standalone databases. We need more storage and servers to house the current enormous amount of data that will only increase in the future. A single database is unable to accommodate this data fortune.

Achieving prompt query data response time

Even though a DBMS has to accommodate enormous amounts of data, the experience and response time that's expected by customers and users do not allow DBMS downtime to organize the data little by little. How to retrieve the requested data from the data lake will be one of the top issues.

Querying and storing fragmented data types

Furthermore, the relational data structure has become one part of various data types. Documents, JSON, graphs, and key-value pairs are all attracting people's attention. This is reasonable since all of them come from varying business scenarios that involve keeping the world moving smoothly and efficiently.

All these new changes and requirements will bring necessary challenges and needs to databases and their operation and maintenance.

You may have been aware of or even already encountered some of these expectations in your professional experience. If you are just stepping into the professional world, you are bound to encounter these expectations, no matter your future industry. This is because the role of the database administrator has changed. More precisely, it has evolved, and the next section will tell you how.

The evolving role of the DBAs

These changes in industry needs have reshaped the role of the DBA as we know it. While the role of DBAs is crucial within any organization, whether it is a technology business or not, its importance has been growing at a speed that is directly correlated to the digital technology adoption rate. They are constantly looking for ways to optimize their database management systems and are the primary strategy designers to counter data spikes and ensure data safety and data availability.

They've been long considered to be key guardians of the vital strategic asset of data. This responsibility is not narrow in scope as it includes many other duties. DBAs must ensure their organizations can meet their data needs, that databases perform at optimal levels and function properly, and that, in case of any issues, they are called upon to recover the data.

Over the past decade, their responsibilities have also been reshaped thanks to new data-producing devices (smartphones and IoT devices, for example) that continue to drive data growth, thus ultimately increasing the number of database instances under management, as well as a wider array of database management systems. More recent developments have even seen DBAs increasingly involved in application development, making them emerging key influencers in the overall data management infrastructure.

In the next few sections, we will look into the most common and pressing challenges that DBMSs are facing today, and for which a DBA should be prepared.

Overwhelming traffic load increase

Ever since the introduction of the iPhone, mobile phones have gained an increasingly important role in our lives, allowing us to do more than place and receive phone calls while on the go. We now shop, order food, book our vacations, do our banking, hunt for jobs, consume entertainment, and connect with our family and friends thanks to the little devices in our pockets. While this interconnectivity gave rise to multiple new industries and business models (think sharing economy and calling an Uber), they all have one thing in common: data. The amount of data we consume and produce has ballooned to levels that were inconceivable just 15 years ago.

With the advent of the internet, it has become the norm for successful websites or business services that support apps to be receiving visits that reach well into the billions every week.

Sales days such as Cyber Monday in North America or 11/11 (also known as Singles Day) in China (the largest shopping festival in the world) are excellent examples of traditional retail enterprises that adapted to the digital world. Now, they must contend with new needs to successfully achieve their business goals. In cases such as these, retailers are looking to drive traffic to their pages or online stores. But what happens if they succeed and their database clusters are put under incredible pressure? The question becomes a technical one, with DBAs and R&D teams wondering if their database cluster will be able to handle the visitors' traffic.

Microservice architecture for frontend services

To deal with a large number of visitors, the monolithic architecture has since been phased out and officially became history. Instead, microservices architecture has become the new favorite.

A microservices architecture integrates an application as an *ensemble* of weakly related services. In other words, this results in an application being built as a set of independent components running the process as a service, performing a part of the whole system. Lightweight APIs are how these components communicate, with each service allowing for deployment, updates, and scaling according to specific business requirements as they are run independently.

Cloud-native disrupts delivery and stale deployment practices

The advent of the cloud has brought deep and significant changes, including overturning the way to host, deliver, and start up software.

One of the major changes that can be attributed to the advent of the cloud is the conceptual advance it brought by breaking the barrier between hardware and software. All our media, emails, and the digits of our bank accounts are spread across thousands of servers controlled by hundreds of companies. This is even more impressive if we consider that, not even 20 years ago, the internet was in its inception stages, and only used by early adopters or academics that knew how to search a directory or operate an FTP file.

In a sense, the cloud is the natural result of the stars aligning and all the right conditions being met. If we look back, we can see how the success of the cloud was thanks to the wider adoption of broadband internet, the higher penetration rate of smartphones, allowing constant internet connectivity, and all the other innovations that made data centers easier to build and maintain. This is one of the rare instances where enterprise and consumer innovation seem to be advancing at a comparable pace. From a consumer angle, we can already see how physical storage will soon be unnecessary thanks to the internet, while for business needs, we now find offerings that allow us to run computing tasks on third-party servers – even for free.

In the perennial pursuit of flexibility, many enterprises are now moving their technologies to the cloud because of the scalability and affordability it brings. Being flexible can arguably be interpreted as being adaptable, which is exactly what executives would be after to be able to respond to industry or broader market changes. Plus, it opens the door for startups to sell their product and services directly on the cloud. It also allows them to build, manage, and deploy their applications anywhere with freedom and flexibility.

Considering the significant potential opportunities offered by the cloud, some organizations have already started to adopt a cloud-first strategy, which simply means including or moving to a cloud-based solution at the expense of a strategy built around in-house data centers. This new IT trend is going to move the databases to the cloud as a **Database-as-a-Service (DBaaS)**.

Considering the numerous and significant changes and requirements that businesses and services face in their quest for digital transformation, to keep up the pace with their relative industries, we can easily understand the drive behind companies' motivation to change the way they store, query, and manage data from their databases. The following diagram shows how databases are used to store, query, and manage data:

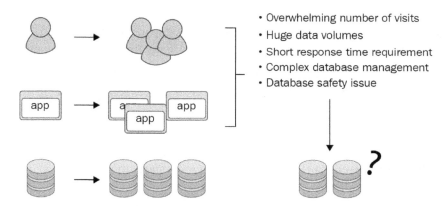

Figure 1.1 – Database challenges flow

As you can see, the databases on the right are marked with a question mark. This represents two things: what are the possibilities, and what are the directions that you can undertake in your role as a database professional to be prepared for them?

In the next section, you will be introduced to the opportunities and future directions that you should be aware of when it comes to databases. Not only can they give you an advantage in your profession, but they can also help you chart your career if you keep them in mind when it's time to make decisions about your professional development.

The opportunities and future directions for DBMSs

Let's review the opportunities, as well as future directions, that DBMSs are headed in. In the next few subsections, you will encounter topics ranging from database security to industry novelties such as DBaaS.

Database safety

Database safety has been one of the key focus areas for DBMSs. On the one hand, database vendors strive to deliver and iterate on existing solutions to solve database issues.

Cloud vendors are committed to protecting the data and applications that exist in the cloud infrastructure. The internet, software, load balancers, and all the components of the data transmission flow are seeing their safety measures being upgraded one by one.

Considering this ongoing improvement process, the natural question that arises is this: how can we achieve the seamless integration that's needed between the projects that are developed in different languages and various databases?

To answer this question and the necessary challenges that come with dealing with such important questions, we are seeing an increasingly significant number of resources being dedicated to both leading enterprises and promising new start-up ventures.

More than two-thirds of CIOs are concerned about the constraints that could emerge because of cloud providers. It is for these reasons that open source databases are becoming the go-to solution.

Data security has not only become paramount for enterprises but can be the determinant between survival or being forgotten forever as another firm that went out of business. If you think about ransomware and how it is increasingly widespread, you may be able to understand how open source technology empowers organizations to defend themselves against such risks. Open source allows organizations to be in total control of their security needs by giving them complete access to source code, as well as the flexibility that comes with being able to configure and extend the software as they see fit.

There is certainly a counter-argument to the criticism about the security of open source that was prevalent years ago. Rapid adoption by enterprises seems to be settling the argument in favor of open source. No company will remain untouched by the power of open source database progress.

SQL, NoSQL, and NewSQL

When SQL is brought up in a conversation, people immediately think about the good old relational database, which has been supporting higher-level services for the past couple of decades.

Unfortunately, the relational database has since started to show its age and is now considered by many as not adequate to meet the new requirements that businesses must nowadays respond to. This has caused industry giants in the database field to take aggressive actions to reshape their product offerings or deliver new solutions.

NoSQL is one such example. It was the initiator of the non-relational database, which provides a mechanism for storing and retrieving data modeled in a non-relational fashion, such as key-value pairs, graphs, documents, or wide columns. Nevertheless, many NoSQL products compromise consistency in favor of availability and partition tolerance. Without transaction and SQL's standard advantages, NoSQL databases gain the high availability and elastic scale-out that's necessary to respond to the vital concerns of the new era. The success of Couchbase, HBase, MongoDB, and others all stand as clear evidence in support of this thesis. NoSQL databases also sometimes emphasize that they are *Not Only SQL* and that they do recognize the value of the traditional SQL database. This type of appreciation has led to NoSQL databases gradually adopting some of the benefits of mainstream SQL products.

NewSQL can be defined as a type of **relational database management system** (**RDBMS**) looking to make NoSQL systems scalable for **online transaction processing** (**OLTP**) tasks, all while keeping the ACID qualities of a traditional database system.

The discussion is still ongoing both in academia and in the industry, with the definition being regarded as fluid and evolving. An excellent resource is the paper *What's Really New with NewSQL?* (`https://dl.acm.org/doi/10.1145/3003665.3003674http s://dl.acm.org/doi/10.1145/3003665.3003674`), which set out to categorize the databases according to their architecture and functions.

All the databases shouting out they are one of the NewSQL products are seeking a nice balance between capability, availability, and partition tolerance (CAP theorem). But which products belong to NewSQL?

New architecture

Among the opportunities for DBMSs that are currently available and stated to bring significant changes to the industry in the short to medium term, new database architectures certainly merit consideration. This is where databases are effectively designed from an entirely new code base, thus leaving behind any of the architectural baggage of legacy systems – a *clean slate* of sorts that allows for near endless possibilities, as new databases are being conceptualized and built to meet the needs of the new era.

Embracing a transparent sharding middleware

A transparent sharding middleware splits a database into multiple shards that are stored across a cluster of a single-node DBMS instance, just as Apache ShardingSphere does. Sharding middleware exists to allow a user – or in this case, an organization – to split a database into multiple shards to be stored across multiple single-node DBMS instances, such as Apache ShardingSphere. This section will help you understand what data sharding is. Database administrators are constantly looking for ways to optimize their database management systems. When data input spikes, you must have strategies in place to handle it. One of the best techniques for this is to split the data into separate rows and columns, and such examples include data sharding or partitioning. The following sections will introduce you to, or refresh, these concepts and the difference between them.

Data sharding

When a large database table is split into multiple small tables, **shards** are created. The newly created tables are called shards or partitions. These shards are stored across multiple nodes to work efficiently, improving scalability and performance. This type of scalability is known as **horizontal scalability**. Sharding eventually helps database administrators such as yourself utilize computing resources in the most efficient way possible and is collectively known as **database optimization**.

Optimizing computing resources is one key benefit. More critical is that the network can scan fewer rows and respond to queries on the user side much faster than going through one colossal database.

Data partitioning

When we talk about partitioning, it may sound confusing. The reason for your potential confusion is completely normal as data partitioning is often mistakenly thought about when it comes to data sharding.

Partitioning refers to a database that has been broken down into different subsets but is still stored within a single database. This single database is sometimes referred to as the database instance. So what is the difference between sharding and partitioning? Both sharding and partitioning include breaking large data sets into smaller ones. But a key difference is that sharding implies that the breakdown of data is spread across multiple computers, either as horizontal partitioning or vertical.

Database-as-a-Service

The DBaaS providers not only provide the remodeled cloud databases but are responsible for maintaining their physical configuration as well. Users do not need to care about where the database is located; the cloud allows the cloud database providers to take care of the physical databases' maintenance and related operations.

NoSQL and NewSQL are unavoidable opportunities, towards which most if not all database vendors are moving and represent the future of DBMSs. Many startups are moving into this space to fill this market gap and deliver services that directly complete with the ones provided by established industry giants.

AI database management platform

The technological developments of the last 10 years are allowing advances in nascent fields such as **machine learning** (**ML**) and **artificial intelligence** (**AI**). Such technologies will eventually impact all aspects of our lives, and enterprises and their databases are no different.

AI database operation and maintenance are poised to become the main growth drivers for the future of DBMSs. The relationship between AI and databases may not seem to be evident at first; while AI has become a sort of *buzzword* these days, database management has remained automatic, platform-based, and observed while requiring intensive human interaction.

When AI technology is eventually integrated into databases' operation and maintenance work, new avenues will be opened. The historical experience of the previous operations that were performed during database management tasks will be machine-learned, and databases empowered by AI will be able to provide suggestions and specific actions to manage, operate, maintain, and protect database clusters.

Furthermore, AI database management platforms will also be able to contact the monitoring and warning system, or even undertake some pressing operations to avoid significant production accidents. Productivity improvement and headcount optimization reduction are always central concerns of enterprises.

Database migration

When it comes to database migration, there is some good news and some bad news. In the spirit of optimism about the future, let's consider the good news first: we have new database candidates, such as all the NewSQL and NoSQL offerings that have hit the market recently.

When it comes to the bad news, it'll be necessary to be able to deliver data migration at the lowest price.

In this old-to-new process, data migration and database selection occupy an important part of peoples' minds. Many enterprises choose to stick with stale database architecture to avoid any negative effect on production and the instability that could be caused by new databases.

Additionally, legacy and complicated IT systems contribute significantly to discouraging risk-taking, and confidence in performing data migration. In such cases, many database vendors or database service companies will offer to develop new products for this bulky work and insert themselves into this market to get a piece of the billion dollars' worth pie that is the database industry.

To recap, some of the main opportunities for DBMSs in the future include database security, leveraging new database architectures, considering embracing data sharding or DBaaS, and fully mastering database migration.

Before moving on to the next section, there is one last thing you have probably already thought of at some point in your career. There are still concerns during this old-to-new-database transition period, such as the following:

- On-premises versus the cloud
- The lowest cost to migrate data to new databases
- Increased program refactoring work costs caused by using multiple databases

The following diagram illustrates an example of the costs that may be incurred while transitioning from an old to a new database:

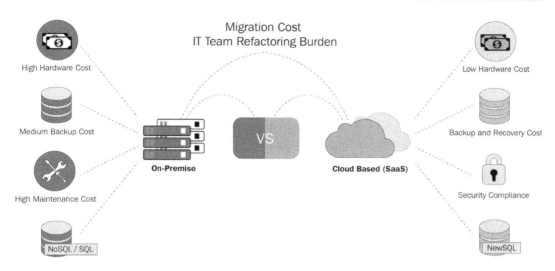

Figure 1.2 – Old-to-new database transition cost

Solving these challenges is not a small feat by any means. There is a multitude of tools and ways that you or an enterprise could employ. The truth is that for most of these solutions, you'd be expected to spend considerable amounts of time and financial resources to succeed as they'd require completely switching database type or vendor, reconfiguring your whole system, or worse, developing custom patches for the databases. Let's not forget that all of these involve risks, such as losing all of your data in the process.

It is for these reasons that we have thought of Apache ShardingSphere. It has been built to be as flexible and unintrusive as possible to make your life easier. You could set it up quickly without having to disturb anything in your system, answer all of the previously mentioned challenges, and set yourself up to be ready for the future developments mentioned in this chapter as well. The next section will give you an introductory overview of what Apache ShardingSphere is and its main concept.

Understanding Apache ShardingSphere

A unified data service platform is the best solution to the bottleneck issues of a peer-to-peer data service model. Apache ShardingSphere is an independent database middleware platform with a supportive ecosystem, positioned as Database Plus, to build a criterion and ecosystem above multi-model databases. The three key elements of Apache ShardingSphere are connect, enhance, and pluggable. We will discuss these concepts in detail in the following sections.

Connect

The basic feature of Apache ShardingSphere is to make it incredibly easy to connect data and applications. Instead of creating a new API to build an entirely new database standard, it chooses to pursue compatibility with existing databases, making you feel as if nothing has changed in your interaction with and among the various original databases.

Its unified database access entry, known as database gateway, enables Apache ShardingSphere to simulate target databases and transparently access databases and their peripheral ecosystems, such as application SDKs, command-line tools, GUIs, monitoring systems, and more. ShardingSphere currently supports many types of database protocols, including MySQL and PostgreSQL protocols.

Connect refers to ShardingSphere's strong database compatibility – that is, building a database-independent connection between data and applications to greatly improve enhanced features.

Enhance

To only connect without including additional service features can already be considered an implementation plan – good or bad as it may be. The result would, in nature, be equivalent to directly connecting databases. Such a plan not only increases network costs but damages performance as well, not to mention the low value you'd get from it.

The primary feature of Apache ShardingSphere is to capture database access entry and provide additional features transparently, such as redirect (sharding, read/write splitting, and shadow databases), transform (data encrypting and masking), authentication (security, auditing, and authority), and governance (circuit breaker, access limitation and analysis, QoS, and observability).

The ongoing trend of database fragmentation makes it impossible to centralize the management of all database features. The additional features of Apache ShardingSphere neither target a single database, nor make up for the shortages of database features; instead, they get rid of the shackle of databases simply serving as storage and give unified services that answer DBMSs' concerns.

Pluggable

The progressive addition of new features has expanded Apache ShardingSphere throughout its development history. To avoid creating a steep learning curve that could discourage prospective new users and developers from integrating Apache ShardingSphere into their database environment, Apache ShardingSphere chose to pursue and ultimately adopt a fully pluggable architecture.

The core value of the ShardingSphere project is not the number of different database access and functions, but its developer-oriented and highly extensible pluggable architecture.

As a developer, you are allowed to create custom features without having to modify the source code of Apache ShardingSphere.

The pluggable architecture of Apache ShardingSphere adopts a microkernel and a three-layer pluggable mode. Apache ShardingSphere's architecture is directed toward top-level APIs, so the kernel cannot be aware of the existence of specific functions. If you don't need a function, all you have to do is delete the dependency – it'll have zero impact on the system.

The following diagram shows how ShardingSphere is built:

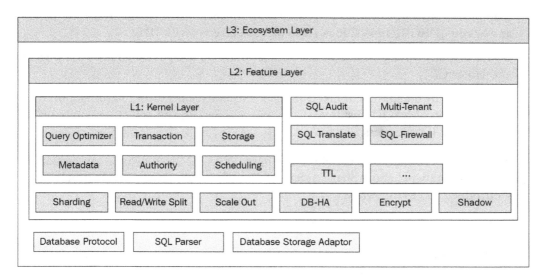

Figure 1.3 – The Apache ShardingSphere ecosystem

As you can see, the three layers are fully independent. Being focused on a plugin-oriented design means that the kernel and feature components fully support ShardingSphere's extensibility, allowing you to build a ShardingSphere instance without it affecting your overall experience if you were to drop (choose to not install) some feature modules, for example.

The architectural possibilities at your disposal

Database middleware requires two things: a driver to access the database and an independent proxy. Since no adaptor of an architecture model is flawless, Apache ShardingSphere chose to develop multiple adaptors.

ShardingSphere-JDBC and ShardingSphere-Proxy are two independent products, but you can choose what we interchangeably refer to as the *hybrid model* or *mixed deployment*, and deploy them together. They both provide dozens of enhanced features that see databases as storage nodes that apply to scenarios such as Java isomorphism, heterogeneous languages, cloud-native, and more.

ShardingSphere-JDBC

Being the predecessor and eventually the first client of the Apache ShardingSphere ecosystem, ShardingSphere-JDBC is a lightweight Java framework that provides extra services at the Java JDBC layer. ShardingSphere-JDBC's flexibility will be very helpful to you for the following reasons:

- It applies to any ORM framework based on JDBC, such as JPA, Hibernate, Mybatis, and Spring JDBC Template. It can also be used directly with JDBC.

- It supports any third-party database connection pool, such as DBCP, C3P0, BoneCP, and HikariCP.

- It supports any database that meets JDBC's standards. Currently, ShardingSphere JDBC supports MySQL, PostgreSQL, Oracle, SQLServer, and any other databases that support JDBC access.

You have probably recognized many of the databases and ORM frameworks mentioned in the previous list, but what about ShardingSphere-Proxy's support? The next section will quickly introduce you to the proxy.

ShardingSphere-Proxy

ShardingSphere-Proxy was the second client to join the Apache ShardingSphere ecosystem. A transparent database proxy, ShardingSphere-Proxy provides a database server that encapsulates the database binary protocol to support heterogeneous languages. The proxy is as follows:

- Transparent to applications; it can be used directly as MySQL/PostgreSQL.

- Applicable to any kind of client that is compatible with the MySQL/PostgreSQL protocol.

The following diagram illustrates ShardingSphere-Proxy's topography:

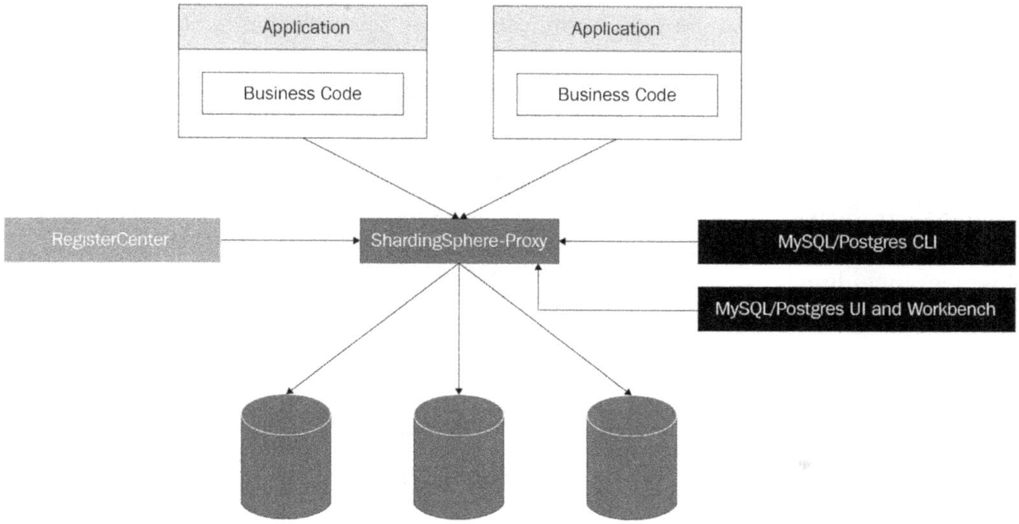

Figure 1.4 – ShardingSphere-Proxy's topography

As you can see, ShardingSphere-Proxy is not intrusive and easily fits into your system, offering you great flexibility.

You may be wondering what the differences between the two clients are and how they compare. The next section will offer you a quick comparison. For a more in-depth comparison, please refer to *Chapter 5*, *Exploring ShardingSphere Adaptors*.

Comparing ShardingSphere-JDBC and ShardingSphere-Proxy

If we consider the simple database middleware projects, different access ends mean different deployment structures. But Apache ShardingSphere is the exception: it supports tons of features. So, with the increasing demand for big data computing and resources, different deployment structures have different resource allocation plans.

ShardingSphere-Proxy has a distributed computing module and can be deployed independently. It applies to applications with multidimensional data calculation, which are less sensitive to delay but consume more computing resources. For more details on the comparison between ShardingSphere-JDBC and ShardingSphere-Proxy, please refer to *Chapter 5*, *Exploring ShardingSphere Adaptors*, or https://shardingsphere. apache.org/document/current/en/overview/.

Hybrid deployment

Adopting a decentralized architecture, ShardingSphere-JDBC applies to Java-based high-performing and lightweight OLTP applications. On the other hand, ShardingSphere-Proxy provides static entry and comprehensive language support and is suitable for OLAP applications, as well as managing and operating sharding databases.

This results in a ShardingSphere ecosystem that consists of multiple endpoints. Thanks to a unified sharding strategy and the hybrid integration of ShardingSphere-JDBC and ShardingSphere-Proxy, a multi-scenario-compatible application system can be built with ShardingSphere. The following diagram introduces an example topography of a hybrid deployment including both ShardingSphere-JDBC and ShardingSphere-Proxy:

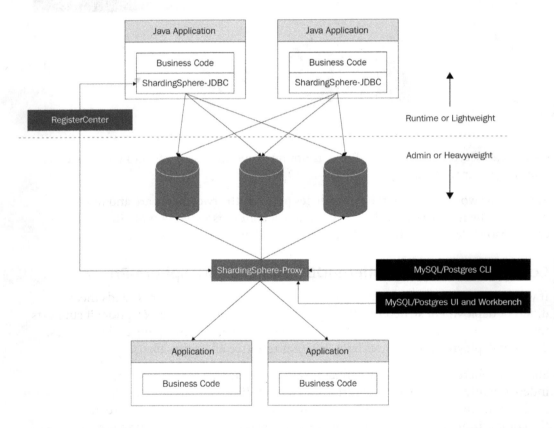

Figure 1.5 – ShardingSphere hybrid deployment topography

When you deploy ShardingSphere-JDBC and ShardingSphere-Proxy together, as shown in the preceding diagram, a hybrid computing capability will be obtained. This allows you to adjust the system architecture to optimally suit your needs.

Summary

In this chapter, you learned about the evolution of DBMSs, the industry pain points, and the new requirements presented by the industry when it comes to databases. This implies that the role of the DBA has to keep up and adapt or, if you will, evolve. If you are reading this book, you are on the right track as you're probably aware of the significant changes that are taking place in the database field and want to be ahead of the curve.

Being ahead of the curve and answering the challenges that databases are facing today has been – and is – our community's driver for developing Apache ShardingSphere. The last section of this chapter gave you a brief introduction to the ShardingSphere clients and how it is built, but this is just the beginning. You have 11 more chapters ahead of you, and by the time you complete them, you not only will master ShardingSphere – you will have acquired a new tool, expanded your skillset, and placed yourself ahead of the curve of upcoming database changes.

By the time this book is published, Apache ShardingSphere will still probably be a unique product in the industry that's aiming to achieve a blue ocean strategy by building Database Plus standards, rather than drowning in the red ocean of distributed databases. A unified database service platform is the only solution to fragmented database tech stacks. Remember, ShardingSphere was born to solve this problem and build the criteria and the ecosystem above multi-model databases.

The next chapter will start your deep dive into Apache ShardingSphere by giving you an architectural overview of the project.

2
Architectural Overview of Apache ShardingSphere

In this chapter, we will introduce you to Apache ShardingSphere's architecture, which also serves as a gateway to a deeper understanding of what a distributed database is. A thorough introduction of the architecture is essential for you to develop a fundamental understanding of how ShardingSphere is built so that you can better use it in your production environment.

We will guide you through some new concepts that have emerged in the database field, such as **Database Mesh**, while also sharing with you the driving development concept of our community—**Database Plus**.

We will start by introducing the typical architecture of a distributed database and then proceed by looking at *layers*, as ShardingSphere is built on a *three-layer* logic.

The first layer includes the kernel with critical features working in the background to make sure everything runs properly on your database. These critical features are a transaction engine, a query optimizer, distributed governance, a storage engine, an authority engine, and a scheduling engine.

Next, we will introduce the second layer, which is the one you will probably be most interested in. In this section, we will give an overview of the features that you can choose to use from this layer, and what their functionality is. Example features are data sharding, elastic scaling, a shadow database, and **application performance monitoring (APM)**.

Finally, we will introduce the third layer, which arguably makes ShardingSphere unique: the pluggable ecosystem layer.

After reading this chapter, you will have a fundamental understanding of how Apache ShardingSphere is built and will also have had an overview of each feature.

In this chapter, we will cover the following topics:

- What is a distributed database architecture?
- The **Structured Query Language** (**SQL**)-based load-balancing layer.
- Apache ShardingSphere and Database Mesh.
- Solving database pain points with Database Plus.
- An architecture inspired by the Database Plus concept.
- Deployment architecture.
- Plugin platform.

What is a distributed database architecture?

A distributed database consists of three inseparable layers—that is, a **load-balancing layer**, a **compute layer**, and a **storage layer**. A distributed database is a type of database in which data is stored across various physical locations. The data stored in said database is not only physically distributed across locations but is also structured and related to other data according to a predetermined logic. The following diagram illustrates the three-layer architecture of distributed database clusters:

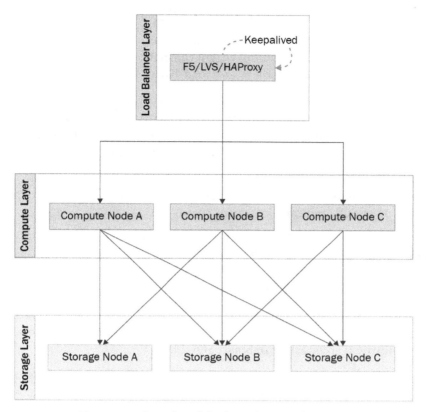

Figure 2.1 – Distributed database cluster architecture

Under the distributed database architecture with storage separated from compute, the stateful storage layer designated for data persistence and push-down computing cannot be expanded as desired. To avoid data loss, it's of great importance to keep multiple copies of data and to adopt a dynamic migration solution to scale out.

The stateless computing layer, on the other hand, is created for distributed query plan generation, distributed transaction, and distributed aggregate calculation, allowing users to horizontally scale computing capabilities. Since computing nodes are scalable, we decided to build the load balancer in front of the database clusters. Naturally, it becomes the centralized point of entry.

In this section, we looked at the concept of distributed databases. This will be useful for understanding the following sections, as Apache ShardingSphere sets out to present a solution allowing the transformation of virtually any of the **relational database management systems** (**RDBMSs**) mentioned in *Chapter 1, The Evolution of DBMSs, DBAs, and the Role of Apache ShardingSphere*, into a distributed database system.

The SQL-based load-balancing layer

The network load-balancer layer is now mature enough, as it can distribute and process requests via protocol header identification, weight calculation, and current limiting. Nevertheless, the industry currently still lacks a load-balancer layer suitable for SQL.

Currently, there isn't a load-balancing layer that can understand SQL, which means that a lack of SQL interpretation causes unsatisfactory granularity of request dispatching in any database system. Developing a Smart SQL **Load Balancer** (**LB**) solution capable of understanding SQL is the best solution to supplement the database load-balancer layer.

In addition to common load-balancer features such as high performance, traffic governance, **service discovery** (**SD**), and **high availability** (**HA**), Smart SQL LB also has the capability to parse SQL and calculate query costs.

When it really understands SQL characteristics and query costs, its next move is to label computing nodes and even to tag storage nodes as well. You can use custom labels, of course, such as the following: SELECT && $cost<3, UPDATE && $transaction=true && $cost<10, and (SELECT && GROUP) || $cost>300.

Smart SQL LB can match the executed SQL with a predefined label and distribute the request to the right computing node or storage node. The following diagram shows the features and deployment architecture of Smart SQL LB:

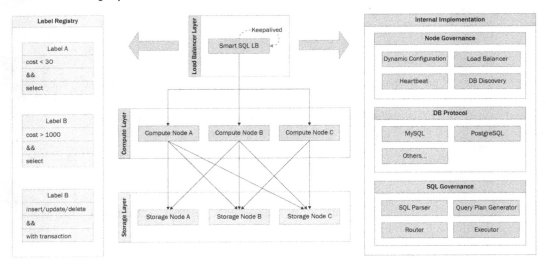

Figure 2.2 – Smart SQL LB deployment architecture

In *Figure 2.2*, the middle part displays the distributed database cluster architecture with Smart SQL LB, in which the load-balancer layer maintains its HA via **organization and management (O&M)** methods such as Keepalive and **Virtual Internet Protocol (VIP)**.

The right side of the preceding diagram shows the kernel design of Smart SQL LB. By simulating the target database protocol, it implements the proxy of the load-balancer layer to make accessing Smart SQL LB consistent with directly accessing the target database (MySQL, PostgreSQL, and so on). In addition to the previously mentioned capability to understand SQL, basic features such as dynamic configuration, heartbeat monitors, database discovery, and load balancers are included in node management.

On the left side of the diagram is a configuration example of user-defined labels for the computing node and the storage node. The labels are stored in the Registry Center of Smart SQL LB, and thus the compute layer and the storage layer only need to process the request.

The two layers do not bother with load-balancer-layer capabilities. Thus, the expected return time of a SQL request is quite close to the time the request is sent to the corresponding labeled compute node or storage node. Such improvement solves two pain points in large-scale database clusters, as outlined here:

- Greatly improved system **quality of service (QoS)** makes the whole cluster run in a much more stable fashion and minimizes the probability of single-node performance waster.

- It effectively isolates transaction computing, analytical computing, and other operations with finer granularity, making cluster resource allocation more reasonable; it's also convenient to customize node hardware resources by following the label description.

In the next section, we will check out how to improve performance and availability thanks to the integration of a load balancer with Sidecar.

Sidecar improves performance and availability

We all agree that performance and availability are at a system's core, rather than fancy-sounding or powerful extra features. Unfortunately, adding the new load-balancer layer may produce a change in both performance and availability.

An additional load balancer increases the network hop count and therefore affects performance to some extent. Concurrently, we also need to build another **Virtual Internet Protocol (VIP)** or *another load balancer designed for the load balancer* to fix the HA problem of the load-balancing layer per se, but this method would end up making the system even more complicated. Sidecar is the silver bullet used to solve the problem caused by adding a new load-balancer layer.

A **Sidecar model** refers to the addition of a load balancer on each application server. Its life cycle is consistent with that of the application itself: when an application starts, so does its load balancer; when an application is destroyed, it will disappear as well.

Now that every application is equipped with its own load balancer, the HA of the load balancer is also ensured. This method also minimizes performance waste because the load balancer and the application are deployed in the same physical container, which means the **Remote Procedure Call** (**RPC**) is converted into **Inter-Process Communication** (**IPC**).

Along with better performance and HA, Sidecar also has applications loosely coupled to its database **software development kit** (**SDK**), and therefore **operations** (**Ops**) teams can freely upgrade Sidecar or databases since the design shields the business application from perceiving the upgrade.

Since Sidecar is processed independently of the application, as long as the canary release of Sidecar is guaranteed, it will be completely transparent for application developers. Unlike class libraries dispatched to application programs, Sidecar can more effectively unify online application versions.

The popular Service Mesh concept actually uses Sidecar as its dashboard to process east-west and north-south traffic in the system, and also leverages the Control Panel to issue instructions to the dashboard in order to control traffic or complete transparent upgrades.

The biggest challenge of the model is its costly deployment and administration because it is necessary to deploy each application with Sidecar. Due to the high demand for deployment quantity, it's critical to measure its resource occupancy and capacity.

> **Tip**
> Being extremely lightweight is a requisite for a good Sidecar application.

Database Mesh innovates the cloud-native database development path

Kubernetes is a good way to lower Sidecar deployment and management costs. It can put a load balancer and the application image in a Pod or use a **DaemonSet** to simplify the deployment process. After each Pod starts running, Sidecar can be understood as an inseparable part of the operating system. The application accesses the database through `localhost`.

For applications, there will always be a database with unlimited capacity that never crashes. Kubernetes is already the de facto standard for cloud-native operating systems, so deploying Sidecar with Kubernetes in the cloud is definitely accepted by the cloud industry.

Along with the success of Service Mesh, the concept of service-oriented, cloud-native, and programmable traffic has revolutionized the Service Cloud market.

Currently, cloud-native databases still focus on cloud-native data storage, yet there is no universally acclaimed architecture innovation like Service Mesh that allows adaptors to be smoothly delivered through the internet. Database Mesh, supported by the Sidecar solution consisting of Kubernetes and Smart SQL LB, can definitely innovate cloud-native databases and, at the same time, it can better parse SQL.

The three core features of Database Mesh are the load-balancer layer, which can understand SQL, programmable traffic, and cloud-native. The Database Mesh architecture is shown in the following diagram:

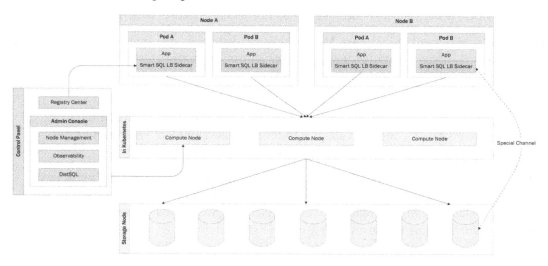

Figure 2.3 – Database Mesh architecture

The Control Panel manages the load balancing layer, the compute layer, and the storage layer; it can even govern all database traffic. The panel involves the Registry Center and Admin Console.

The Registry Center is used for the distributed coordination of SD, metadata storage (such as label definition, and the mapping information of compute nodes and storage nodes), and storage of the operating status of each component in a cluster.

In the Admin Console, node management and observability are the key capabilities of cloud-native distributed databases. They manage the resources of the entire cluster through cloud management and telemetry.

Apart from resource control, the Admin Console allows SQL commands to operate cluster configuration. The SQL used to control distributed clusters is different from the SQL used for database execution, so we created a new type of SQL named **Distributed SQL** (**DistSQL**) to manage distributed clusters.

DistSQL is a supplementary SQL. Together with some necessary features such as traffic management and observability, it can also manage labels (such as defining labels and changing the matching relationship between compute nodes and storage nodes, and so on) to change cluster traffic direction. DistSQL is powerful and flexible enough to enable the Control Panel to dynamically modify traffic control and the router of the entire cluster via programming. It's very similar to the Service Mesh dashboard, but Database Mesh is placed at a different layer. Service Mesh, relying on network traffic, does not need to understand SQL semantics, while Database Mesh supplements database traffic cloud-native control.

Actually, the dashboard of Database Mesh is the load-balancer layer that understands SQL, because it receives commands sent by the Control Panel and executes operations such as current limiting, fusing, and label-based routing.

Database Mesh can completely isolate an environment from another, so operation engineers only need to change the network configuration of the dashboard into that of a distributed database, and then adapt it to the right development environment, test environment, or production environment.

Developers only need to develop localhost-oriented database services. It's not necessary to perceive distributed databases after all. Thanks to the cloud-native service functionality provided by Database Mesh, engineers can completely ignore specific database network addresses, which greatly improves their workflow efficiency.

Apache ShardingSphere and Database Mesh

Although Database Mesh and Apache ShardingSphere may sound similar in some aspects, they are not the same. For example, in contrast to Database Mesh, ShardingSphere's Smart SQL LB doesn't intrude into compute or storage nodes, ultimately making it truly adaptable to any kind of database.

However, a combination of Database Mesh and Apache ShardingSphere can improve interaction performance through a private protocol.

Smart SQL LB can generate an **abstract syntax tree** (**AST**) via the SQL parser. So, in the future, starting from version 5.1.0, Apache ShardingSphere will open a private protocol: when it receives SQL requests, it can also receive an AST concurrently to improve its performance in the appropriate manner. For example, apart from SQL parsing, in some scenarios such as single-shard routing, it's feasible to identify SQL features and directly access backend database storage nodes without Apache ShardingSphere.

Enhanced Smart SQL LB functions plus a practicable private protocol will make Apache ShardingSphere and Smart SQL LB even more compatible. Eventually, they will present an integrated Database Mesh solution.

Solving database pain points with Database Plus

The kernel design concept of Apache ShardingSphere is Database Plus. The database industry has been expanding rapidly over the last few years, with new players and new solutions being offered with the intent to solve the increasingly widening gap created by the development of internet-related industries.

Some notable examples include MongoDB, PostgreSQL, Apache Hive, and Presto. Their popularity shows us that database fragmentation is an increasingly widespread issue in the database industry, and the relatively recent pouring of venture-capital support has only exacerbated its growth and development. According to *DB-Engines* (`https://db-engines.com/en/`), there are more than 350 databases ranked, with many more that didn't even make the list. According to **Carnegie Mellon University's** (**CMU's**) *Database of Databases* (`https://dbdb.io/`), there currently are 792 different noteworthy **database management systems** (**DBMSs**).

Such a large number of DBMSs is telling of the wide spectrum of possible requirements different businesses may have when it comes to choosing their DBMS. However, a coin has two sides, and this *database boom* is bound to create issues, as described here:

- The upper application layer needs to contact each different type of database with different database dialects. Preserving existing connection pools while adding new ones becomes an issue.

- Aggregating distributed data among separate databases becomes increasingly challenging.

- Another challenge is how to apply the same needs to fragmented databases— such as encryption, for example.

- **Database administrators'** (**DBAs'**) responsibilities will increase due to maintaining and operating different types of databases in the production environment, which may cause inefficiencies at scale.

We thought of Database Plus in order to build a standardized layer and ecosystem that would be positioned above the fragmented databases, providing unified operation services and hiding database differentiation. This would create an environment where applications are required to only communicate with a standardized service.

Moreover, additional functions are possible to be included with the flow from applications to databases. The resulting mechanism sees this layer—that is, ShardingSphere—acquiring abilities to hijack the traffic, parse all requests, modify the content, and reroute these queries to anticipated target databases.

Because of Database Plus' basic but important idea, ShardingSphere can provide sharding, data encryption, a database gateway, a shadow database, and more.

As mentioned in *Chapter 1, The Evolution of DBMSs, DBAs, and the Role of Apache ShardingSphere*, the core elements of the Database Plus concept are **link**, **enhance**, and **pluggable**, as outlined here:

- **Link** means connecting applications and databases. In some cases, applications do not need to be aware of the databases' variable differences, even if a given type of database is removed or added to the bottom of this architecture. Applications are not required to be adapted to this change since they would have been communicating with ShardingSphere from the beginning.

- **Enhance** means to improve the database capability thanks to ShardingSphere. Why would a user prefer to interact through ShardingSphere rather than the original databases?

  That is simply because, admittedly, the **link** element alone would not be persuasive enough for users to consider using ShardingSphere. Therefore, the **enhance** element is necessary for ShardingSphere and can provide users with additional and valuable features such as sharding, encryption, and authentication, while at the same time linking databases with applications.

- When it comes to the **pluggable** element, it is enough to consider that users have diversified needs and issues. They want noticeable care for their needs, which means that user-defined rules and configurations or customization are necessary. However, product sellers or vendors stand in the opposite direction. They prefer to provide standardized products to avoid extra labor, development, or customization costs. The motivating drive has always been to find an answer to the following question: *Is there a way to break this deadlock?* The community went to work and made ShardingSphere provide **application programming interfaces (APIs)** for most functionalities. Its kernel workflow basically interacts with these APIs to run. Hence, no matter what the specific implementation is, ShardingSphere can work well while presenting both official and default implementations for each function for an out-of-the-box option together with infinite customization possibilities.

Furthermore, all these functions are pluggable with each other. For example, if you are not looking to implement sharding in your environment for your use case, you can skip its plugin configuration and just provide a data-encryption setting file, and you will get an encrypted database.

Conversely, if you plan to use sharding and encryption functions together, just tell ShardingSphere about this expectation, and it will package these two configurations to build an encrypted sharding database for you. The following diagram presents an overview of the gap created between applications and fragmented databases by new industry needs. As we introduced earlier, the *database boom* has created a fragmented market that still cannot meet the diversified service requirements brought by a plethora of newly developed applications. Failing to meet fast-changing user needs is what is creating a gap in the industry:

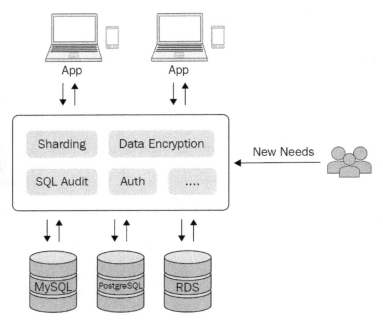

Figure 2.4 – New industry creating a gap between applications and fragmented databases

When it comes to the benefits of Database Plus, we could summarize them with the following points:

- Standardized layer to hide different usages for various databases.
- Noticeably reduce the effect of varying database replacement.
- Supply enhanced functions to solve these annoying problems.
- Assemble different feature plugins for your specific cases.
- Allow users to utilize their customized implementations for most kernel phases.

This innovative idea kicked off with the Apache ShardingSphere 5.x releases. Previously, ShardingSphere defined itself as a sharding middleware layer to help users shard their databases. At that time, it was a lightweight driver, totally different from the present orientation.

Users were vocal in sharing their hopes of ShardingSphere being able to support more valuable features in addition to sharding. To respond to the community's expectations, other excellent features have been added to the development schedules.

On the other hand, the simple combination of various features made the architecture difficult to maintain and hard to keep up with in a sustainable manner. The endeavor to solve these issues required a long-term undertaking. We struggled with these needs and the initial chaos of the project, to finally be able to initiate the Database Plus concept bearing the three characteristic core elements we mentioned previously.

This iterative process sets ShardingSphere apart from other similar sharding products. Citus and Vitess are popular and admittedly excellent products to scale out PostgreSQL or MySQL. Currently, Citus and Vitess are concentrating on sharding and other relevant features, which makes them similar to the older versions of ShardingSphere. The ShardingSphere of today is tracing a new and innovative path, which is bound to drive the database industry toward new heights.

An architecture inspired by the Database Plus concept

Based on the Database Plus concept, ShardingSphere is built according to the architecture that we will introduce you to over the next few sections. We will learn more about this project from both feature architecture and deployment architecture perspectives. Multiple perspectives will allow you to understand how to use the Database Plus concept and deploy it in a production environment.

Feature architecture

Feature architecture elaborates more on the clients, features, and supported databases. It's a catalog of each available component—a dictionary including its clients, functions, layers, and supported databases.

As previously introduced in *Chapter 1, The Evolution of DBMSs, DBAs, and the Role of Apache ShardingSphere*, all components—such as data sharding, data encryption, and all of ShardingSphere's features, including clients—are optional and pluggable. As graphically represented in the following diagram, once you select ShardingSphere for your environment, your next step is to select one or more clients, features, or databases to assemble a database solution that fits your needs:

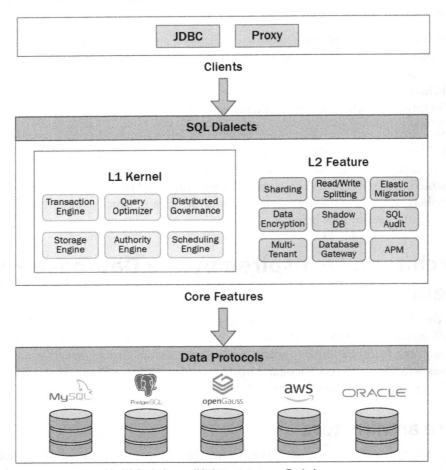

Figure 2.5 – ShardingSphere's feature architecture

Now that we have introduced an overall idea of the architecture, the following sections will allow you to have a deeper understanding of the clients, features, and database protocols in order to make informed decisions.

Clients

The top layer is composed of **ShardingSphere-Proxy** and **ShardingSphere-JDBC**, which are two products or clients for you to choose from, and they have respectively suitable usages and scenarios. You can choose to deploy one of them or both together. If you remember, we covered these two clients and their deployment patterns in *Chapter 1*, *The Evolution of DBMSs, DBAs, and the Role of Apache ShardingSphere*. If you need a quick refresher, just remember that ShardingSphere-JDBC is a lightweight Java framework, while ShardingSphere-Proxy is a transparent database proxy.

Introduction to the feature layer

Let's now look at the layer of the architecture that includes the features of Apache ShardingSphere. We will first look at the kernel layer and its corresponding features and then proceed to **layer 2 (L2)**—the function layer and its features. This is useful for gaining an overview of where features are located before we jump to the deployment architectures in the following section.

L1 kernel layer

If you remember, we first introduced this concept and its definition in *Chapter 1*, *The Evolution of DBMSs, DBAs, and the Role of Apache ShardingSphere*, in *Figure 1.3*. This layer is part of the *feature* layer but focuses on the kernel part, which indicates internal features that are critical but have no direct importance in solving general concerns. They run in the background and support the L2 *function* layer. Sometimes, their quality determines the performance of these features located in the L2 layer, and in other cases, they notably influence the performance of all clients. Therefore, massive amounts of effort and time were dedicated to their improvement. The following sections will provide you with an understanding of the components of the kernel layer and their functionalities.

Transaction engine

Database transactions are required to satisfy **atomicity, consistency, isolation, and durability (ACID)** conditions. When considering a single database, transactions can only access and control a single database's resources and are referred to as local transactions. These are natively supported by the overwhelming majority of mature relational databases.

On the other hand, we also have distributed transactions. An increasing number of transactions require the inclusion of multiple service accesses and the respective database resources in the same transaction, thanks to distributed application situations based on microservices.

The transaction engine is designed to deal with local transactions, **eXtended architecture (XA)**-distributed transactions, and **basically available, soft State, eventually consistent (BASE)**-distributed transactions.

Query optimizer

The query optimizer attempts to analyze SQL queries and determine the most efficient way to execute a query by considering multiple possible query plans. In a distributed database system, a query optimizer determines the data volume to acquire, transit, and calculate from multiple shards.

Once SQL queries are submitted to the database server and parsed by the parser, it is the right time to pass them to the query optimizer. Since the structure of data and query is not one-to-one, the gap between them is likely to cause inefficient data acquisition.

All the operators in the query optimizer will be used to calculate the cost for each query thanks to the metadata and optimization rules, to subsequently generate various execution plans. The final target of the query optimizer is to find the best execution plan to process a given query in minimal time. Nevertheless, a good enough execution plan is more acceptable. This is because the optimization process to find the best way still takes some time, which might slow performance.

Distributed governance

A distributed system requires individual components to have unified management ability. Distributed governance essentially refers to the ability to manage the state of database storage nodes and the driver or proxy computing nodes, while being able to detect any updates in the distributed environment and synchronize them among all computing nodes online in real time. Furthermore, it will collect the metadata and information of this distributed system and provide advice for this database cluster. By means of **machine learning (ML)** technology, this distributed governance can learn the history of your database and calculate more beneficial database personalized recommendations.

Additionally, circuit-breaker and traffic-limiting features provided by distributed governance ensure the whole database cluster can serve applications continuously, efficiently, and fluently.

Storage engine

A storage engine concentrates on how to store data in a certain structure. From the previous architecture shown in *Figure 2.5*, you will see various databases such as MySQL, PostgreSQL, Oracle, Server, **Amazon Relational Database Service (Amazon RDS)**, and so on. Clearly, they are all databases, which means they have the capabilities to store data and calculate queries. In fact, they can be regarded as storage solutions, and the responsibility of the storage engine is to communicate with these storage solutions.

How do you differentiate database middleware from distributed databases? Some people will give the answer that the type of storage and the connection to them are the main factors. If this distributed system is built from scratch (storage, computing nodes, and clients), that means it is born as a distributed database for the new era's needs. Conversely, when the storage is a DBMS and the computing nodes utilize SQL to retrieve data from storage, that is viewed as a database middleware. In other words, the key is the presence or lack of native storage.

When it comes to the storage engine of ShardingSphere, it includes and connects different storage, which implies that the storage could be a DMBS or a **key-value** (**KV**) storage. A specific function is taken charge of by the corresponding engine so that other modules are not influenced by the modification or changes of other parts. This design also creates chances for storage replacement.

Authority engine

An authority engine provides user-authentication and privilege-control capabilities for distributed systems. Through the different configuration levels supported by an authority engine, you can use password encryption, schema-level privilege control, table-level privilege control, and more fine-grained privilege customization to provide different degrees of security for your data.

At the same time, a privilege engine is also the basis of data auditing. With the help of a privilege engine, you can easily complete the configuration of audit privileges, thereby limiting the scope of data queries for some users or limiting the number of rows affected by certain SQL.

In addition to the way files are configured, the authority engine also supports the Registry Center or integrates with the user's existing authority system through an API so that architects can flexibly choose a technical solution and customize the authority system that is most suitable for them.

Scheduling engine

A scheduling engine is designed for job scheduling; it is powered by Apache ShardingSphere ElasticJob, which supports an elastic-schedule, resource-assign, job-governance, and job-open ecosystem.

Its features include support for data sharding and HA in a distributed system, scale-out to improve throughput and efficiency, and flexible and scalable job processing thanks to resource allocation. Additional noteworthy highlights include self-diagnosis and recovery when the distributed environment becomes unstable.

It is used for elastic migration, data encryption, data synchronization, online **Data Definition Language** (**DDL**), shard splitting and merging, and so on.

L2 function layer

The following sections will provide you with an understanding of the components and functionalities of the function layer. The features of the L2 function layer are supported by background-running components of the L1 function layer that were introduced in the *L1 kernel layer* section.

Sharding

When we talk about sharding, we refer to splitting data that is stored in one database in a way that allows it to be stored in multiple databases or tables, in order to improve performance and data availability. Sharding can usually be divided into database sharding and table sharding.

Database sharding can effectively reduce visits to a single database and thereby reduce database pressure. Although table sharding cannot reduce database pressure, it can reduce the data amount of a single table to avoid query performance decrease caused by an increase in index depth. At the same time, sharding can also convert distributed transactions into local transactions, avoiding the complexity of distributed transactions.

The sharding module's goal is to provide you with a transparent sharding feature and allow you to use the sharding database cluster like a native database.

Read/write splitting

When we talk about read/write splitting, we refer to splitting the database into a primary database and a replica database. Thanks to this split, the primary database will handle transactions' insertions, deletions, and updates, while the replica database will handle queries.

Recently, databases' throughput has increasingly been facing a **transactions-per-second (TPS)** bottleneck. For applications with massive concurrent read but smaller simultaneous write, read/write splitting can effectively avoid row locks caused by data updates and greatly improve the query performance of the entire system.

A primary database with many replica databases is able to improve processing capacity thanks to an even distribution of queries across multiple data replicas.

This also means that if we were to expand this setup with multiple primary databases with multiple replica databases, we can enhance not only throughput but also availability. Such advantages are no small feat, and you shouldn't overlook this type of configuration. Remember that under this configuration, the system can still run independently on the occurrence of any database being down or having its disk physically destroyed. The goal of the read/write splitting module is to provide you with a transparent read/write splitting feature so that you can use the primary replica database cluster like a native database.

Elastic migration

Elastic migration is a common solution for migrating data from a database to Apache ShardingSphere or scaling data in Apache ShardingSphere. It's a type of scale-out/-in or horizontal scaling, but not scale-up/-down or vertical scaling.

Considering a hypothetical situation where our business data is growing quickly, then according to widespread database professionals' belief, the backend database could be the bottleneck. How could we prevent or resolve this? Elastic migration could help.

Just add more database instances and configure more shards in Apache ShardingSphere, then migration will be scheduled and a scaling job will be created to do the migration. The scaling job includes four phases: a preparation phase, an inventory phase, an incremental phase, and a switching phase. You will learn more about these phases and the elastic-scaling workflow in *Chapter 3*, *Key Features and Use Cases – Your Distributed Database Essentials*, under the *An introduction to elastic scaling* section. Database connectivity and permissions checks, as well as recording the position of logs, will be done in the preparation phase. Inventory data on current time migration will be done in the inventory phase.

In the event that data is still changing during the inventory phase, the scaling job will synchronize these data changes thanks to the **change data capture** (**CDC**) function. The CDC function is based on database replication protocols or **write-ahead logging** (**WAL**) logs. This function will be performed in the incremental phase. After the incremental phase is completed, the configuration could be switched by `register-center` and `config-center` so that the new shards will be online during the switching phase.

Data encryption

Data security is of paramount importance, both for internet enterprises and ones belonging to more traditional sectors, and data security has always been an important and sensitive topic.

At the core of data-security notions, we find data encryption, but what is it exactly? It is the transformation of sensitive information into private data through the use of encryption algorithms and rules.

If you're thinking about what some potential examples could be, just think of data involving clients' sensitive information, which requires data encryption to be compliant with consumer data protection regulations, for example (such as **identification** (**ID**) number, phone number, card number, client number, and other personal information).

The goal of the data-encryption module is to provide a secure and transparent data encryption solution.

Shadow DB

This is a suitable solution for the current popular microservice application architecture. Businesses require multiple services to be in coordination, therefore the stress test of a single service can no longer represent the real scenario.

The industry usually chooses online full-link stress testing—that is, performing stress testing in a production environment. To ensure the reliability and integrity of production data to prevent data pollution, data isolation has become a key and difficult point.

The Shadow DB function is ShardingSphere's solution for isolating the pressure-test data at the database level in a full-link pressure-test scenario.

APM

ShardingSphere's APM feature provides metrics and tracing data to a framework or server for implementing the observability of ShardingSphere. It's based on **Byte Buddy**, which is a runtime code generator used by Java agents, and it's developed this way to support zero intrusion into other modules' code and decoupling it from core functions.

The APM module can be released independently. Currently, it supports exporting metrics data to Prometheus, and Grafana can easily visualize this data through charts. It supports exporting tracing data to different popular **open source software** (**OSS**) such as Zipkin, Jaeger, OpenTelemetry, and SkyWalking. It also supports the OpenTracing APIs at the code level. The metrics data includes information about connections, requests, parsing, routing, transactions, and part of the metadata in ShardingSphere.

The tracing data includes the time elapse of parsing, routing, and execution of SQL information alongside the request. The useful data exported by the APM module can provide you with a simple way to analyze ShardingSphere runtime performance. The APM module can also support monitoring the user application using the ShardingSphere-JDBC framework as its backend data access.

Deployment architecture

ShardingSphere provides many practical deployment patterns to solve your cases, but this section will introduce the necessary components with an illustrative and simple one.

Applications are the external visitors, and a computing node, a proxy, is in charge of receiving traffic, parsing SQL, and calculating and scheduling distributed tasks. The registry will persist the metadata, rules, configurations, and cluster status. All the databases will become storage nodes persisting the data and running some calculation jobs.

The following diagram shows one of the three possible deployment architectures you can use with ShardingSphere. These are Proxy, **Java database connectivity** (**JDBC**), or hybrid, as mentioned in *Chapter 1, The Evolution of DBMSs, DBAs, and the Role of Apache ShardingSphere*, under the *The architecture possibilities at your disposal* section:

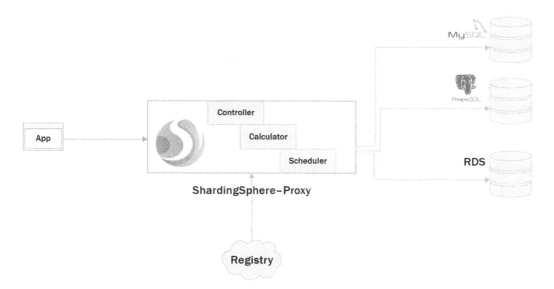

Figure 2.6 – ShardingSphere-Proxy deployment architecture

If you have some knowledge of Kubernetes architecture, you will find that it shares similarities with ShardingSphere's architecture. The following diagram gives us an example of Kubernetes architecture, which we can compare with *Figure 2.6*:

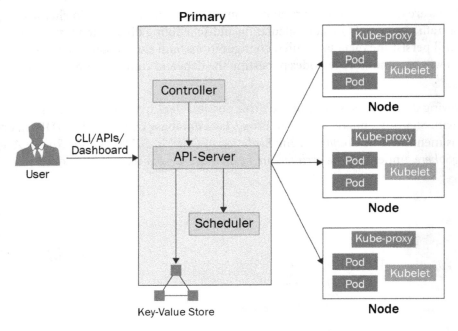

Figure 2.7 – Kubernetes architecture example

As you can see from the preceding diagram, the primary node and proxy work alike. The primary node controls the nodes as represented in *Figure 2.7*, just as—similarly—proxy visits these databases. etcd, which is a distributed key-value store of a distributed system's critical data, is used to manage the cluster state, and all the primary nodes connect to it. The registry also needs all the proxies to register to it.

These two diagrams will hopefully allow you to understand the similarities between ShardingSphere-Proxy and typical Kubernetes architectures. This will be useful as we proceed to the next section where we will learn about the plugin platform of Apache ShardingSphere.

Plugin platform

Engineers may choose ShardingSphere-JDBC or ShardingSphere-Proxy, but when they use Apache ShardingSphere, they actually call the same kernel engine to process SQL. Though Apache ShardingSphere with its complex structure contains many features, its microkernel architecture is extremely clear and lightweight.

Microkernel ecosystem

The core process of using Apache ShardingSphere is very similar to that of using a database, but ShardingSphere contains more core plugins for users, together with extension points. An overview of the kernel architecture is shown in the following diagram:

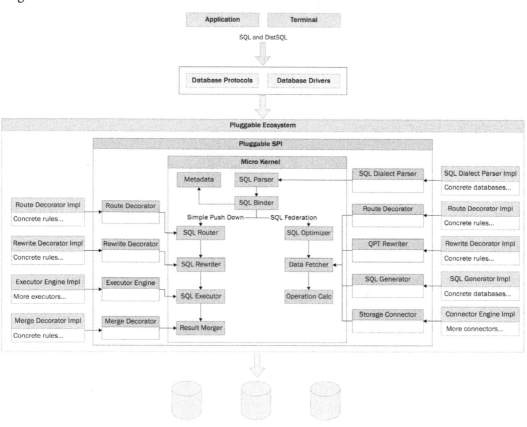

Figure 2.8 – Kernel architecture overview

The preceding diagram shows the three layers of the architecture—namely, the innermost microkernel, the pluggable **service provider interfaces** (**SPIs**) in the middle, and the pluggable ecosystem placed at the outermost layer.

The microkernel processing workflow involves two standard modules, SQL Parser and SQL Binder. These two modules are developed to identify specific SQL characteristics and then, based on the results, the SQL execution workflow is divided into a simple push-down engine and a SQL federation engine.

The pluggable SPI is the abstract top-level interface of Apache ShardingSphere's core process. The microkernel does not work for rule implementation, and it just calls a class registered in the system that implements an interface step by step. Along with the SQL executor, all SPIs use the decorator design pattern to support feature combinations.

A pluggable ecosystem allows a developer to use pluggable SPIs to combine any desired features. The core features of Apache ShardingSphere (for example, data sharding, read-write splitting, and data encryption) are pluggable SPIs or components of the pluggable ecosystem.

During the microkernel process, SQL Parser converts the SQL you input into an AST Node through the standard lexer and parser processes, and eventually into a SQL statement to extract features, which is actually the core input of the Apache ShardingSphere kernel processer.

SQL statement embodies the original SQL, while SQL binder combines metadata and SQL statement to supplement wildcards and other missing parts in the SQL, to generate a complete AST node that conforms to the database table structure.

The SQL parser analyzes basic information—for example, it checks whether SQL contains related queries and subqueries. SQL binder analyzes the relations between logical tables and physical databases to determine the possibility of cross-database-source operation for the SQL request. When the full SQL can be pushed down to the database storage node after operations such as modifying a logical table or executing information completion, it's time to adopt Simple Push Down Engine to ensure maximum compatibility with SQL. Otherwise, if the SQL involves cross-database association and a cross-database subquery, then SQL Federation Engine is used to achieve better system performance during the process of operating distributed table association.

Simple Push Down Engine

For Apache ShardingSphere, Simple Push Down Engine is an old feature, and it is applicable to scenarios where database native computing and storage capabilities need to be fully reused to maximize SQL compatibility and stabilize query responses. It is perfectly compatible with an application system based on the *Share Everything* architecture model.

The following diagram clearly illustrates the architecture of Simple Push Down Engine. Following the two standardized preprocesses, *SQL Parser* and *SQL Binder*, *SQL Router* extracts the key fields in SQLStatement (for example, shard key) and matches them with specific DistSQL rules configured by the user to calculate the routing data source:

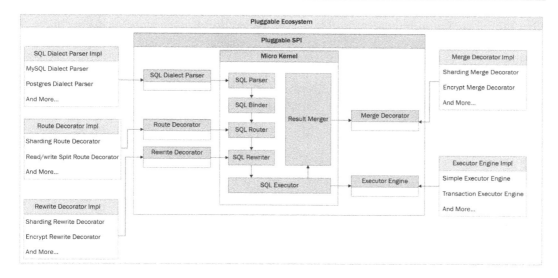

Figure 2.9 – Simple Push Down Engine architecture

When the data source is clarified, *SQL Rewriter* rewrites the SQL so that it can be directly pushed down to the database for execution in a distributed scenario. Execution types include logical table name replacement, column supplementation, and aggregate function correction.

SQL Executor checks the transaction status and selects a suitable execution engine to execute the request; then, it sends the rewritten SQL in groups to the data source of the routing result concurrently.

Finally, when execution results of SQL are completed, *Result Merger* automatically aggregates multiple result sets or rewrites them one more time.

Having acquired an understanding of the Simple Push Down Engine architecture and how it works with SQL, the next step will be understanding ShardingSphere's SQL Federation Engine.

SQL Federation Engine

SQL Federation Engine is a newly developed engine in Apache ShardingSphere, but its development iteration is quite frequent now. The engine is suitable for cross-database associated queries and sub-queries, and its architecture is shown in the following diagram.

It can fully support a multi-dimensional elastic query system applicable to the *Share-Nothing* distributed architecture model. Apache ShardingSphere will continue to improve the query optimizer model, trying to migrate more SQLs from Simple Push Down Engine to SQL Federation Engine.

Before jumping right into the details of how SQL Federation Engine works and how it is built, take a minute to have a look at the following diagram we prepared to give you an overview of SQL Federation Engine's architecture:

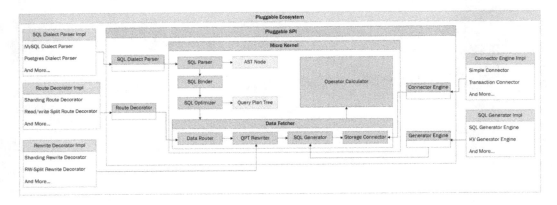

Figure 2.10 – SQL Federation Engine architecture

Compared with Simple Push Down Engine, it is more similar to the core process of databases. The major difference between SQL Federation Engine and Simple Push Down Engine is the *SQL optimizer*, which optimizes the *AST Node* by leveraging a **rule-based optimizer (RBO)** and **cost-based optimizer (CBO)** to generate a *Query Plan Tree*.

Therefore, to obtain data from the storage node, it does not rely on the original SQL but instead can leverage the *query plan tree*, regenerate a new SQL that can be executed on a single data node, and later send this to the storage node according to the routing result.

Before the final SQL is generated, **Query Plan Tree Rewriter (QPT Rewriter)** modifies the table or the column according to the DistSQL rules configured by the user. It is worth noting that the target data source does not have to be consistent with the SQL dialect input by the user. Instead, it can use the query plan tree to generate a new SQL dialect suitable for storage nodes, such as other database dialects, and even KV.

In the end, it returns the execution result to the compute node of Apache ShardingSphere, and *Operator Calculator* finishes the final aggregate calculation in the memory until the result set is finally returned.

Both Simple Push Down Engine and SQL Federation Engine showcase how core extension points of Apache ShardingSphere are open to enhanced features. As long as they master the design process of the microkernel architecture, developers can easily add custom functions by implementing core extension points.

Summary

With this chapter, we set out to understand how Apache ShardingSphere is built from an architecture point of view, in order to establish a foundational understanding of how its ecosystem works.

First, we thought it would be necessary to share with you our motivations and some of the recent developments brought to the database industry by Kubernetes and Database Mesh. Having an overview of these concepts allows you to gain an understanding of our Database Plus development concept, Apache ShardingSphere's three-layer architecture, and its components. We then learned about the features, the importance of SQL parsing, and how Apache ShardingSphere is transitioning.

In the next chapter, we will introduce you to potential use cases for Apache ShardingSphere in a professional environment, while introducing its features in a detailed manner.

Section 2: Apache ShardingSphere Architecture, Installation, and Configuration

On completion of Part 2, you will be able to fully grasp the potential of the multitude of features that the ShardingSphere ecosystem offers. Once you have developed a thorough understanding of this potential, you will learn how to successfully install, customize, and deploy a customized version of Apache ShardingSphere.

This section comprises the following chapters:

- *Chapter 3, Key Features and Use Cases – Your Distributed Database Essentials*
- *Chapter 4, Key Features and Use Cases – Focusing on Performance and Security*
- *Chapter 5, Exploring ShardingSphere Adaptors*
- *Chapter 6, ShardingSphere-Proxy Installation and Startup*
- *Chapter 7, ShardingSphere-JDBC Installation and Startup*

3
Key Features and Use Cases – Your Distributed Database Essentials

If you have completed the previous chapter, then you will have already acquired an understanding of how Apache ShardingSphere is built, its kernel, and its architecture.

You may also remember how the previous chapter introduced some features that are located in the third layer (L3) of ShardingSphere's architecture. This chapter will introduce you to the most important features of Apache ShardingSphere: the features that will allow you to create a distributed database and perform data sharding – including how they are integrated into its architecture and how they work.

After completing this chapter, you will have expanded your knowledge of the architecture that you gathered in the previous chapter, not only with new knowledge about each specific feature but also the key use cases for each feature.

As you may recall, in *Chapter 1*, *The Evolution of DBMSs, DBAs, and the Role of Apache ShardingSphere*, we introduced you to the main pain points that database professionals, as well as businesses at large, are facing as a result of digital transformation and the app economy. Reading this chapter will allow you to connect each feature and its respective use case with a corresponding real industry pain point.

After reading this chapter, you will have mastered the features of Apache ShardingSphere that are necessary for a distributed database, their use cases, and the pain points that they address.

These features include read/write splitting, high availability, observability, and data encryption. The common themes that tie these features together are performance and security; implementing them will allow you to improve your system's performance and security.

In this chapter, we will cover the following topics:

- Distributed database solutions
- Understanding data sharding
- Understanding SQL optimization
- Overview and characteristics of distributed transactions
- An introduction to elastic scaling
- Read/write splitting

Distributed database solutions

In the context of the current internet era, business data is growing at a very high pace. Faced with storage and access to massive amounts of data, the traditional relational database solution of single-node storage is experiencing significant challenges. It is difficult to meet massive data in terms of its performance, availability, operation, and maintenance cost:

- **Performance**: Relational databases mostly use B+ tree indexes. When the amount of data is too large, the increase in the index depth will increase the amount of disk access I/O, resulting in the decline of query performance.

- **Availability**: As application services are stateless, they can realize low-cost local capacity expansion making the database become the bottleneck of the whole system. It has become increasingly difficult for traditional single data nodes or primary-secondary architectures to bear the pressure of the whole system. For these reasons, databases' availability has become increasingly important, to the point of becoming any given system's key.

- **Operation and maintenance costs**: When the amount of data in the database instance increases to a certain level, the operation and maintenance costs will greatly increase. The cost in terms of time lost for data backup and recovery will eventually grow exponentially, and in a way that is directly proportional to the amount of data being managed. In some cases, when relational databases cannot meet the storage and access requirements of massive data, some users store data in the original NoSQL that supports distribution. However, NoSQL has incompatibility problems with SQL and imperfect support for transactions, so it cannot completely replace the relational database – and the core position of the relational database is still unshakable.

Facing this rapid expansion in the amounts of data to be stored and managed, the common industry practice is to use the data fragmentation scheme to keep relational databases' data volume of each table below the threshold by splitting the data, to meet the storage and access requirements of massive data.

The key method to achieve this is via data sharding, which is not to be confused with data partitioning. In the next section, we will introduce you to data sharding, why sharding might be the answer to your data storage issues, and how Apache ShardingSphere supports and implements this technology.

Understanding data sharding

Data sharding refers to splitting the data in a single database into multiple databases or tables according to a certain dimension, to improve performance and availability.

Data sharding is not to be confused with data partitioning, which is about dividing the data into sub-groups while keeping it stored in a single database. Many other opinions and ideas are floating around in academia and on the internet about this, but rest assured that the number of databases where the data is stored represents the main difference that you should be aware of when distinguishing between sharding and partitioning.

According to the granularity of data sharding, we can divide data sharding into two common forms – database shards and table shards:

- **Database shards** are partitions of data in a database, with each shard being stored in a different database instance.

- **Table shards** are the smaller pieces that used to be part of a single table and are now spread across multiple databases.

Additionally, database shards can efficiently disperse the number of visits to a single point of the database and alleviate the pressure on the database. Although table shards cannot alleviate database pressure, they can convert distributed transactions into local transactions, avoiding the complexity that's brought by distributed transactions.

According to the data sharding methodology, we can divide data sharding into vertical sharding and horizontal sharding. Let's dive deeper into the characteristics of both types of sharding.

Understanding vertical sharding

Vertical sharding refers to splitting the data according to your requirements, and its core concept is dedicated to special databases. Before splitting, a database consists of multiple tables, and each table corresponds to different businesses. After splitting, tables are classified according to the business and distributed to different databases. This helps disperse the access pressure to different databases.

Vertical sharding usually requires architecture and design adjustments, and it cannot solve the single-point bottleneck of the database. If the amount of data in a single table is still too large after vertical sharding, it needs to be processed further through horizontal sharding.

The following diagram shows the necessary scheme you would utilize to vertically shard user and order tables to different databases:

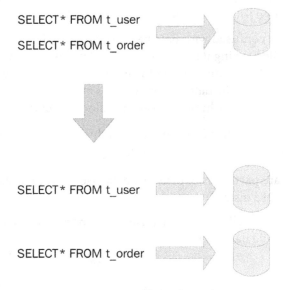

Figure 3.1 – Vertical sharding scheme

You now understand what vertical sharding is, but what about horizontal sharding? For the answer to this question, move on to the following section, where we will introduce you to horizontal sharding – Apache ShardingSphere's first-ever feature.

Understanding horizontal sharding

Horizontal sharding refers to dispersing data into multiple databases or tables according to certain rules through a field (or several fields). Here, each shard contains only part of the data. The following diagram shows the scheme for horizontally sharding user tables to different databases and tables according to primary key sharding.

Theoretically, horizontal sharding breaks through the single database point bottleneck and expands relatively freely. It's a standard solution for data sharding. The following diagram shows a schematic representation of the horizontal sharding concept:

Figure 3.2 – Horizontal sharding scheme

Although data sharding solves performance, availability, operation, and maintenance issues, it also introduces new problems. After sharding data, it becomes very difficult for application development engineers and database administrators to operate the database if they wish to update, move, edit, or reconfigure the data. At the same time, many SQLs that can run correctly in the single-node database may not run correctly in the database after sharding.

Based on these problems caused by data sharding, the ShardingSphere data sharding module provides users with a transparent sharding function, allowing users to use the database cluster after horizontal sharding, just like using the native database (https://medium.com/@jeeyoungk/how-sharding-works-b4dec46b3f6).

Data sharding key points

To reduce the cost of using the sharding function and realize data sharding transparency, ShardingSphere has introduced some core concepts, including table, data node, sharding, row expression, and distributed primary key. Now, let's understand each of these concepts in detail.

Table

Table is the key concept surrounding data sharding. ShardingSphere provides a variety of table types, including logical table, real table, binding table, broadcast table, and single table, to meet the needs of data sharding in different scenarios:

- **Logical table**: This refers to the logical name of the horizontal split database (table) with the same structure and is the logical identification of the table in SQL. For example, if the order data is divided into 20 tables according to the primary key's module, which are t_order_0 to t_order_19, respectively, then the logical table name of the order table is t_order.

- **Real table**: The real table refers to the physical table that exists in the horizontally split database – that is, the logical representation of t_order_0 to t_order_9.

- **Binding table**: A binding table refers to the main and sub-tables with consistent sharding rules. Remember that you must utilize the sharding key to associate multiple tables with a query. Otherwise, Cartesian product association or cross-database association will occur, thus affecting your query efficiency.

- **Broadcast table**: The table that exists in all sharding data sources is called the broadcast table. The table's structure and its data are completely consistent in each database. Broadcast tables are suitable for scenarios where the amount of data is small and needs to be associated with massive data tables, such as dictionary tables.

- **Single table**: On the other hand, a table that only exists in all of the sharding data sources is called a single table. Single tables apply to tables with a small amount of data and without sharding.

Now that you understand the *table* concept, let's move on to the next data sharding concept: data node.

Data node

Data node is the smallest unit of data sharding. It is the mapping relationship between the logical table and the real table. It is composed of the data source's name and the real table.

When configuring multiple data nodes, they need to be separated by commas; for example, `ds_0.t_order_0`, `ds_0.t_order_1`, `ds_1.t_order_0`, `ds_1.t_order_1`. Users can configure the nodes freely according to their needs.

If you are to truly understand data nodes and assimilate their composition, you must divide them into sub-categories. This allows you to compartmentalize the terms and knowledge according to a common denominator.

For example, first, you will begin with the sharding key, sharding algorithm, and strategy. As you may have guessed, the common theme here is sharding. Next, you must move on to row expression, followed by the primary key. Once you've completed these steps, you will be ready to jump into the sharding workflow.

Sharding

Sharding mainly includes the core concepts of sharding key, sharding algorithm, and sharding strategy. Let's look at these concepts here:

- The **sharding key** refers to the database field that's used to split the database (table) horizontally. For example, in an order table, if you want to modify the significant figure of its primary key, the order primary key becomes the shard field. ShardingSphere's data sharding function supports sharding according to single or multiple fields.

- The **sharding algorithm** refers to the algorithm that's used to shard data. It supports `=`, `> =`, `< =`, `>`, `<`, `between`, and `in` for sharding. The sharding algorithm can be implemented by developers, or the built-in sharding algorithm syntax of Apache ShardingSphere can be used, with high flexibility.

- The **sharding strategy** includes the sharding key and sharding algorithm. It is the real object for the sharding operation. Because the sharding algorithm is independent, the sharding algorithm is extracted independently.

Row expression

To simplify and integrate the configuration, ShardingSphere provides row expressions to simplify the configuration workload of data nodes and sharding algorithms.

Using row expressions is very simple. You only need to use `${ expression }` or `$->{ expression }` to identify row expressions in the configuration. The row expression uses **Groovy syntax**. It is derived from Apache Groovy, an object-oriented programming language that was made for the Java platform and is Java-syntax-compatible. All operations that **Groovy** supports can be supported by row expression.

Regarding the previous data node example, where we have ds_0.t_order_0, ds_0.t_order_1, ds_1.t_order_0, ds_1.t_order_1, the row expression can be simplified to db${0..1}.t_order${0..1} or db$->{0..1}.t_order$->{0..1}.

Distributed primary key

Most traditional relational databases provide automatic primary key generation technology, such as MySQL's auto-increment key, Oracle's auto-increment sequence, and so on.

After data sharding, it is very difficult for different data nodes to generate a globally unique primary key. ShardingSphere provides built-in distributed primary key generators, such as UUID and SNOWFLAKE. At the same time, it also supports primary key user customization through its interface.

In the previous sections, we introduced you to the world of data sharding and explained and defined its most important components. Now, let's learn how to perform sharding, and what happens behind the scenes to your data and your database once you initiate data sharding.

Sharding workflow

ShardingSphere's data sharding function can be divided into the **Simple Push Down** process and the **SQL Federation** process, based on whether query optimization is carried out.

SQL Parser and SQL Binder are processes that are included in both the Simple Push Down engine and SQL Federation Engine. The following diagram illustrates SQL's flow during sharding, with the arrows at the extreme right and left of the diagram indicating the input and flow of information:

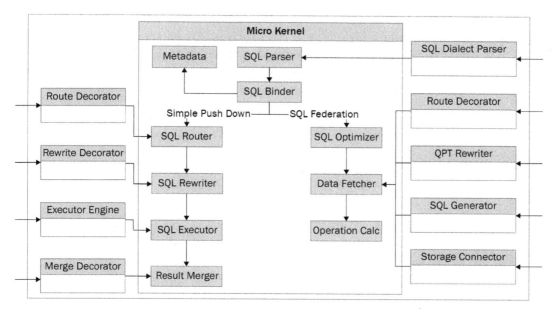

Figure 3.3 – The sharding workflow

SQL Parser is responsible for parsing the original user SQL, which can be divided into lexical analysis and syntax analysis. **Lexical analysis** is responsible for splitting SQL statements into non-separable words; then, the parser, through **syntax analysis**, understands SQL and obtains the **SQL Statement**. You can think of this process as involving lexical analysis, followed by grammar analysis.

The **SQL Statement** includes a table, selection item, sorting item, grouping item, aggregating function, paging information, query criteria, placeholder mark, and other information.

SQL Binder combines metadata and the **SQL Statement** to supplement wildcards and missing parts in SQL, generate a complete parsing context that conforms to the database table structure, and judge whether there are distributed queries across multiple data according to the context information. This helps you decide whether to use the SQL Federation Engine. Now that you have a general understanding of the sharding workflow, let's dive a little deeper into the Simple Push Down engine and SQL Federation Engine process.

Simple Push Down engine

The **Simple Push Down** engine includes processes such as SQL Parser, SQL Binder, SQL Router, SQL Rewriter, SQL Executor, and Result Merger, which are used to process SQL execution in standard sharding scenarios. The following diagram is a more in-depth version of the preceding diagram and presents the processes of the Simple Push Down engine:

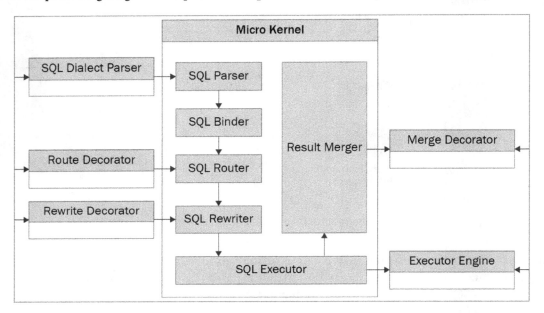

Figure 3.4 – Simple Push Down engine processes

SQL Router matches the sharding strategy of the database and table according to the parsing context and generates the routing context.

Apache ShardingSphere 5.0 supports sharding routing and broadcast routing. SQL with a sharding key can be divided into single-chip routing, multi-chip routing, and range routing according to the sharding key. SQL without a sharding key adopts broadcast routing.

According to the routing context, **SQL Rewriter** is responsible for rewriting the logical SQL that's written by the user into real SQL, which can be executed correctly in the database. SQL rewriting can be divided into correctness rewriting and optimization rewriting. Correctness rewriting includes rewriting the logical table name in the table shard's configuration to the real table name after routing, column supplement, and paging information correction. Optimization rewriting is an effective means of improving performance without affecting query correctness.

SQL Executor is responsible for sending the routed and rewritten real SQL to the underlying data source for execution safely and efficiently through an automated execution engine. SQL Executor pays attention to balancing the consumption that's caused by creating a data source connection and memory occupation. It's expected to maximize the rational use of concurrency to the greatest extent and realize automatic resource control and execution efficiency.

Result Merger is responsible for combining the multiple result datasets that were obtained from various data nodes into a result set and then correctly returning them to the requesting client. ShardingSphere supports five merge types: traversal, sorting, grouping, paging, and aggregation. They can be combined rather than being mutually exclusive. In terms of structure, it can be divided into stream, memory, and decorator merging.

Stream and memory merging are mutually exclusive, while decorator merging can do further processing based on stream and memory merging.

SQL Federation Engine

SQL Federation Engine is a crucial element not only in the implementation of data sharding but in the overall ShardingSphere ecosystem. The following diagram provides an overview of the SQL Federation Engine processes. This will give you a *bird's-eye* view of the whole process before we dive into the details:

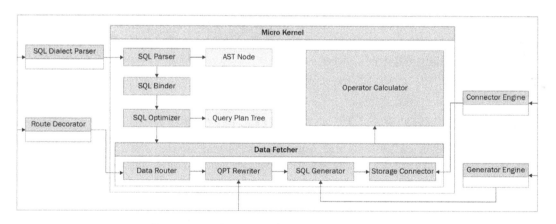

Figure 3.5 – The SQL Federation Engine processes

The SQL Federation Engine includes processes such as **SQL Parser, Abstract Syntax Tree (AST), SQL Binder, SQL Optimizer, Data Fetcher**, and **Operator Calculator,** which are used to process association queries and sub-queries across multiple database instances. The bottom layer uses calculations to optimize the **Rule-Based Optimizer** (**RBO**) and **Cost-Based Optimizer** (**CBO**) based on relational algebra, and query results through the optimal execution plan.

SQL Optimizer is responsible for optimizing the association query and sub-query across multiple database instances, as well as performing rule-based optimization and cost-based optimization to obtain the optimal execution plan.

Data Fetcher is responsible for obtaining data from the storage node according to the SQL that was generated by the optimal execution plan. Data Fetcher also routes, rewrites, and executes the generated SQL.

Leveraging the optimal execution plan and the data obtained from the storage node, the Operator Calculator is responsible for obtaining the query results and returning them to the user.

Why you need sharding

At this point, you have developed an understanding of the concept of data sharding and its key elements. Nevertheless, you may not know what the applications of data sharding are, nor the motivations that may prompt a DBA to implement data sharding. ShardingSphere's data sharding function provides transparent database shards and table shards. You can use the ShardingSphere data sharding function as if you were using a native database. At the same time, ShardingSphere provides a perfect distributed transaction solution, and users can manage distributed transactions in a unified way.

Moreover, combined with scaling, it can realize the elastic expansion of data sharding and ensure that the system can be continuously adjusted with the business, to meet the needs of rapid business growth. Nevertheless, sharding alone is not the solution to all the database issues of today, nor is it the only functionality provided by Apache ShardingSphere. The next section will introduce you to the SQL optimization feature.

Understanding SQL optimization

In a database management system, SQL optimization is crucial. The effect of SQL optimization is directly correlated to the execution efficiency of SQL statements. Therefore, current mainstream relational databases provide some type of powerful SQL optimizer. Based on traditional relational databases, ShardingSphere provides distributed database solutions that include data sharding, read/write splitting, distributed transactions, and other functions.

To meet the associated queries and subqueries across multiple database instances in distributed scenarios, ShardingSphere provides built-in SQL optimization functions through the federation execution engine. This helps achieve the optimal performance of query statements in distributed scenarios.

SQL optimization definition

SQL optimization refers to the equivalent transformation process for query statements and generating efficient physical execution plans using query optimization technology, which usually includes logical optimization and physical optimization:

- **Logical optimization** is RBO and refers to the equivalent transformation rules based on relational algebra to optimize query SQL, including column clipping, predicate pushdown, and other optimization contents.

- **Physical optimization** is CBO and refers to optimizing query SQL based on cost, including table connection mode, table connection order, sorting, and other optimization contents.

Now, let's learn more about RBO and CBO.

RBO

RBO refers to rule-based optimization. The theoretical basis of RBO is relational algebra. It realizes the logical optimization of SQL based on the equivalent transformation rules of relational algebra. It accepts a logical relational algebraic expression before returning the logical relational expression after rule transformation.

The main strategy of ShardingSphere logic optimization is to push down. For example, as shown in the following diagram, pushing the filter condition in an associated query join down to the internal execution of the `join` operation can effectively reduce the amount of data in the `join` operation:

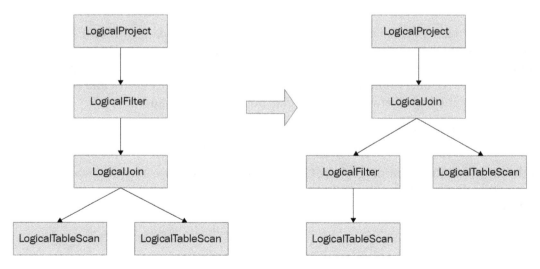

Figure 3.6 – RBO explained

Additionally, considering that the underlying storage of ShardingSphere has computing power, `Filter` and `Tablescan` can be pushed down to the storage layer for execution at the same time, to further improve execution efficiency.

CBO

CBO refers to cost-based optimization. The SQL optimizer is responsible for estimating the cost of the query according to the cost model so that you can select the execution plan with the lowest cost per query. The following diagram shows an example of this:

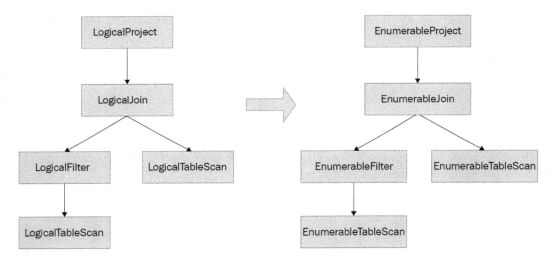

Figure 3.7 – CBO explained

In traditional relational databases, cost estimation is usually based on CPU cost and I/O cost, while SQL optimization in ShardingSphere belongs to distributed query optimization, which makes reducing the number of transmissions and the amount of data as the main goal of query optimization. Therefore, in addition to CPU cost and I/O cost, the communication cost between different data nodes is also considered.

CBO optimization inputs a logical relational algebraic expression and returns the physical relational algebraic expression after optimization.

At the time of writing, ShardingSphere CBO optimization does not introduce statistical information. Instead, it uses the `Calcite VolcanoPlanner` optimizer to realize the transformation from a logical relational algebraic expression to a physical relational algebraic expression.

> **Note**
> Currently, the ShardingSphere SQL optimizer is still an experimental function and needs optimization in terms of performance and memory usage. Therefore, this function is turned off by default. Users can turn this function on by configuring `sql-federation-enabled: true`.

The value of SQL optimization

Through its SQL optimization function, ShardingSphere can support distributed query statements across multiple database instances and provide an efficient query performance guarantee.

At the same time, business R&D personnel no longer need to care about the purpose of SQL, as SQL is automatically optimized across multiple database instances. Also, they can focus on business function development and reduce the functional restrictions at the business level.

You have now gained an understanding of how ShardingSphere optimizes SQL through Simple Push Down and SQL Federation. These are important as SQL represents the way we interact with a database. In the next section, we will learn about one of the features where you will find this information useful: distributed transactions.

Overview and characteristics of distributed transactions

Transactions are important functions for database systems. A **transaction** is a logical unit that operates on a database and contains a collection of operations (generally referred to as SQL). When executing this transaction, it should hold ACID properties:

- **Atomicity**: All the operations in the transaction succeed or fail (be it read, write, update, or data deletion).

- **Consistency**: The status before and after the transaction's execution meets the same constraint. For example, in the classic transfer business, the account sum of the two accounts is equal before and after the transfer.

- **Isolation**: When transactions are executed concurrently, isolation acts as concurrency control. It ensures that the transaction's execution impacts the database in the same way as sequentially executed transactions would.

- **Durability**: By durability, we refer to the insurance that any changes that are made to the data by successfully executed transactions will be saved, even in cases of system failure (such as a power outage).

For a single-point database, it is convenient to support transactions on a physical node. For a distributed database system, a logical transaction may correspond to the operation of multiple physical nodes. Not only should each physical node provide transaction support, but there should also be a coordination mechanism to coordinate the transactions on multiple physical nodes and ensure the correctness of the whole logical transaction.

In the next few sections, you will understand how ShardingSphere supports both CAP and BASE theory through both rigid and flexible transactions. You will learn the difference between the two, as well as the differences between ShardingSphere's local, distributed, and flexible transactions.

Distributed transactions

Based on the CAP and BASE theories, distributed transactions are generally divided into rigid transactions and flexible transactions:

- **Rigid transactions** represent the strong consistency of the data.
- **Flexible transactions** represent the final consistency of the data.

The implementation method for distributed transactions has three roles:

- **Application program (AP)**: This is an application that initiates logical transactions.
- **Transaction manager (TM)**: A logical transaction will have multiple branch transactions to coordinate the execution, commit, and rollback of branch transactions.
- **Resources manager (RM)**: This role executes branch transactions.

Considering distributed transactions' three roles, these are performed while following a two-phase submission:

1. Preparation phase:

   - TM informs RM to perform the relevant operations in advance.
   - RM checks the environment, locks the relevant resources, and executes them. If the execution is successful, it is in the prepared state and not in the completed state, and TM is notified of the execution result.

2. Commit/rollback phase:

   - If TM receives all the results that are returned by RM as successful, it will notify RM to submit the transaction.
   - Otherwise, TM will notify RM to roll back.

> **Note**
>
> Saga transaction and two-phase submission are very similar. The difference is that in the preparation phase, if the execution is successful, there is no prepared state.

If it is a **read-committed (RC)** isolation level, it is already visible to other transactions. When the rollback is executed, a reverse operation will be performed to roll the transactions back to their original states.

> **Note**
>
> Relevant logs will be recorded on both the TM and RM sides to deal with the problem of fault tolerance.

ShardingSphere's support for transactions

As discussed in the previous section, ShardingSphere supports rigid and flexible transactions. It encapsulates a variety of TM implementations through a pluggable architecture and provides you with the interface for begin/commit/rollbacks so that users do not need to care about additional configurations.

When one logical SQL hits multiple storage databases, it ensures the details of the transaction characteristics of multi-branch operations. This helps consistently convey the message across databases. The ShardingSphere transaction architecture is as follows:

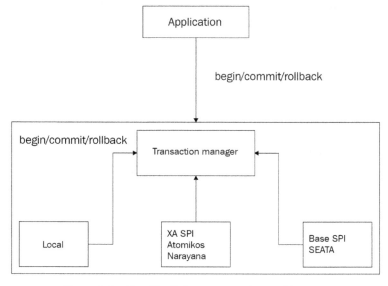

Figure 3.8 – ShardingSphere transaction architecture

ShardingSphere's transaction manager is complete and greatly simplifies your configuration across multiple DBs to help you set up distributed transactions. The next few sections will introduce the different transaction types in detail.

Local transaction

A **local transaction** is based on the transaction of the underlying database. The distributed transaction can support the submission of cross-database transactions and the rollback that's caused by logical errors. Because there is no transaction maintaining the intermediate state, if there are network and hardware-related exceptions during the execution of a transaction, data may be inconsistent.

XA transaction

For XA transactions, two implementations of the open source transaction managers, atomic and Narayana, are integrated to ensure maximum data protection and avoid corruption. Based on the two-phase submission and the XA interface (which we first mentioned in *Chapter 2, Architectural Overview of Apache ShardingSphere*) of the underlying database, you must maintain the transaction log in the intermediate state, which can support distributed transactions.

At the same time, if problems are caused by unresponsive hardware (such as in case of a power outage or crash), the network, and exceptions, the proxy can roll back or commit the transactions in the intermediate state according to the transaction log.

Moreover, you can configure shared storage to store transaction logs (such as using MySQL to store transaction logs). By deploying the cluster mode of multiple proxies, you can increase the performance of proxies and support distributed transactions that include multiple proxies.

Flexible transaction

SEATA's Saga model transaction is integrated to provide a flexible transaction based on compensation. If an exception or error occurs in each branch transaction, the global transaction manager performs the opposite compensation operation and rolls back through the maintained transaction log to achieve the final consistency.

You have now reviewed the definitions and characteristics of all local, distributed, and flexible transactions – but what are the differences between them? Could you pinpoint some key parameters to differentiate them by? The next section will give you the knowledge to do just that.

Transaction modes comparison

The following table compares local, XA, and flexible transactions in terms of business transformation, consistency, isolation, and concurrent performance:

Comparison Parameters	Local Transaction	XA Transaction	Flexible Transaction
Business Transformation	None	None	Set up a seat server instance
Consistency	Unsupported	Support	Eventual consistency
Isolation	Unsupported	Support	Business party guarantee
Concurrent Performance	No effect	Severe recession	Slight decline

Table 3.1 – Transaction modes comparison

If you consider the comparison shown in the preceding table, you will probably be wondering which transaction type will fit which scenario. Depending on the type of scenario you are faced with, you can refer to the following rules of thumb:

- If the business can handle data inconsistencies caused by local transactions that have high-performance requirements, then local transaction mode is recommended.

- If strong data consistency and low concurrency are required, XA transaction mode is an ideal choice.

- If certain data consistency can be sacrificed, the transaction is large, and the concurrency is high, then flexible transaction mode is a good choice.

You now understand the three types of transactions, understand their defining characteristics, and have acquired rules of thumb to determine their suitability in certain scenarios. In the next section, we will introduce you to a scenario that DBAs are increasingly likely to encounter, thus increasing pressure on business support systems.

An introduction to elastic scaling

For a fast-growing business, support systems are generally under pressure, and all the layers of the hardware and software systems may become bottlenecks. In serious cases, there may be problems such as high system load, response delay, or even inability to provide services, thus affecting the user experience and causing the enterprise to incur losses.

From a system perspective, this is a high availability issue. The general solution is simple and direct. The system's pressure issues that are caused by insufficient resources (for example, insufficient storage capacity) can be solved by increasing resources, while excess resources can be solved by reducing resources. This process belongs to capacity expansion and contraction – that is, **elastic scaling**.

There are two types of expansion and contraction schemes in the industry – vertical scaling and horizontal scaling:

- **Vertical scaling** can be achieved by upgrading the single hardware. Affected by Moore's law, which states that the number of transistors on a microchip doubles approximately every 2 years, as well as by hardware costs, with you having to continuously upgrade single hardware, the marginal benefit of this scheme decreases. To solve this problem, the database industry has developed a horizontal scaling scheme.

- **Horizontal scaling** can be achieved by increasing or decreasing ordinary hardware resources. Although horizontal scaling interests hardware, the current state of applications in the computing layer has become relatively mature, and horizontal scaling can be well supported by the **Share-Nothing architecture**. This architecture is a typical architecture for distributed computing since, thanks to its design, each update is satisfied by a single node in the compute cluster.

Currently, elastic scaling is mainly used by the Apache ShardingSphere Proxy solution. Proxy products include a computing layer and storage layer, and both support elastic scaling. The storage layer is the underlying database that's supported by Apache ShardingSphere, such as MySQL and PostgreSQL. The storage layer is stateful, which brings some challenges to elastic scaling.

Glossary

Node: Instances of the computing layer or storage layer component processes can include physical machines, virtual machines, containers, and more.

Cluster: Multiple nodes grouped to provide specific services.

Data migration: Data migration moves data from one storage cluster to another.

Source end: The storage cluster where the original data resides.

Target end: The target storage cluster where the original data will be migrated.

Mastering elastic scaling

So far, you have learned about horizontal and vertical scaling. Elastic scaling, on the other hand, is the ability to automatically and flexibly add, reduce, or even remove compute infrastructure based on the changing requirement patterns dictated by traffic. On the data target side, elastic scaling can be divided into migration scaling and autoscaling. If the target is a new cluster, it is migration scaling. If the target end is a new node of the original cluster, it is auto-scaling. At the time of writing, Apache ShardingSphere supports migration scaling, and auto-scaling is under planning.

In migration scaling, we migrate all the data in the original storage cluster to the new storage cluster, including stock data and incremental data. Data migration to a new storage node takes some time and during the preparation period, the new storage node is unavailable.

How to ensure data correctness, enable the new storage node quickly, enable the new storage node smoothly, and make it so elastic scaling does not affect the operation of the existing system as much as possible are the challenges that are faced by storage layer elastic scaling.

Apache ShardingSphere provides corresponding solutions to these challenges. The implementation method will be described in the following section.

In addition, Apache ShardingSphere also provides good support in the following aspects:

- **Operation convenience**: Elastic scaling can be triggered by DistSQL, and the operation experience is the same as that of SQL. DistSQL is an extended SQL designed by Apache ShardingSphere, which provides a unified capability extension on the upper layer of a traditional database.

- **Degree of freedom of the sharding algorithm**: Different sharding algorithms have different characteristics. Some are conducive to data range queries, while others are conducive to data redistribution. Apache ShardingSphere supports rich types of sharding algorithms. In addition to modulus, hash, range, and time, it also supports user-defined sharding algorithms.

We will discuss these concepts in detail in *Chapter 8, Apache ShardingSphere Advanced Usage – Database Plus and Plugin Platform*. All sharding algorithms support elastic scaling.

The workflow to implement elastic scaling

Having introduced the concepts of vertical and horizontal scaling, let's review how to implement elastic scaling with ShardingSphere. Your general operation process will look like this:

1. Configure automatic switching configuration, data consistency verification, current limiting, and so on.

2. Trigger elastic scaling through DistSQL – for example, modify the sharding count through the `ALTER SHARDING TABLE` rule.

3. View your progress through DistSQL. Receive a failure alarm or success reminder.

Once elastic scaling has been triggered with ShardingSphere, the main workflow of the system will be as follows:

1. **Stock data migration**: The data originally located in the source storage cluster is stock data. Stock data can be extracted directly and then imported into the target storage cluster efficiently.

2. **Incremental data migration**: When data is migrated, the system is still providing services, and new data will enter the source storage cluster. This part of the new data is called incremental data and can be obtained through **change data capture (CDC)** technology, and then imported into the target storage cluster efficiently.

3. **Detection of an incremental data migration progress**: Since incremental data changes dynamically, we need to select the time point where there is no incremental data for subsequent steps to reduce the impact on the existing system.

4. **Read-only mode**: Set the source storage cluster to read-only mode.

5. **Compare the data for consistency**: Compare whether the data in the target end and the source end are consistent. The current implementation is at this stage and the new version will be optimized.

6. **Target storage cluster**: Switch configuration to the new target storage cluster by specifying the data source and rules.

The following diagram shows these steps:

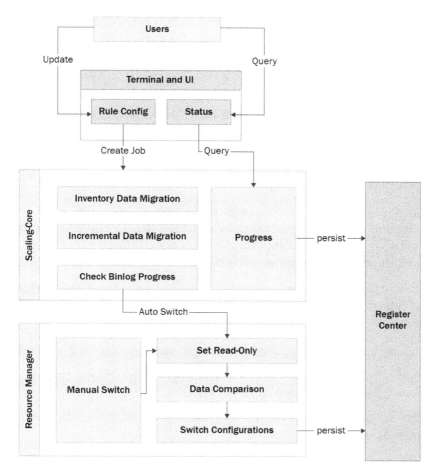

Figure 3.9 – ShardingSphere elastic scaling workflow

The elastic scaling workflow will appear to be straightforward to you as ShardingSphere will handle the necessary steps. As shown in the preceding diagram, the required setup is fairly straightforward, and you only have to interact with the system for the **Rule Configuration** and, if desired, to query the **Status**.

Elastic scaling key points

Scheduling is the basis of elastic scaling. Data migration tasks are divided into multiple parts for parallel execution, clustered migration, automatic recovery after migration exceptions, online encryption and desensitization, and process scheduling at the same time, which are supported by the scheduling system.

Apache ShardingSphere uses the following methods to ensure the correctness of data:

- **Data consistency verification**: This compares whether the data at the source end and the data at the target end are consistent. If the data at both ends is inconsistent, the system will judge that elastic scaling failed and will not switch to a new storage cluster, to ensure that bad data will not go online. The data consistency verification algorithm supports SPI customization.

- **The source storage cluster is set to read-only**: The incremental data changes dynamically. To confirm that the data at both ends is completely consistent, a certain write stop time is required to ensure that the data does not change. The time taken to stop writing is generally determined by the amount of data, the arrangement of the verification process, and the verification algorithm.

Apache ShardingSphere smoothly enables new storage clusters by using an online switching configuration, including data sources and rules. The schema can be kept unchanged, and all proxy cluster (computing layer) nodes can be refreshed to the latest metadata so that the client can continue to be used normally without any operation.

Generally, the most time-consuming part of the process is data migration. The larger the amount of data, the greater the time consumption. Apache ShardingSphere uses the following incremental functions to enable new storage clusters faster:

- The data migration task is divided into multiple parts and executed in parallel.

- The modification operations of the same record are merged (for example, 10 update operations are merged into one update operation).

- We utilize a batch data import to improve speed by dividing the data into batches to not overload the system.

- We have a breakpoint resume transmission feature to ensure the continuity of data transfers in case of an unexpected interruption.

- Cluster migration. This function is under planning and hasn't been completed at the time of writing.

Apache ShardingSphere uses the following methods to minimize the impact on the operation of existing systems:

- **Second-level read-only and SQL recovery**: Regardless of the amount of data, most of it reaches second-level read-only. At the same time, SQL execution is suspended during read-only and is automatically recovered after writing stops, which does not affect the system's availability. At the time of writing, this is under development and hasn't been released yet.

- **Rate limit**: Limit the consumption of system resources via elastic scaling. At the time of writing, this is under development and hasn't been released yet.

These incremental functions are implemented based on the pluggable architecture of Apache ShardingSphere, which was first introduced in *Figure 2.5* in *Chapter 2, Architectural Overview of Apache ShardingSphere*. The pluggable model is divided into three layers: the L1 kernel layer, the L2 function layer, and the L3 ecological layer. Elastic scaling is also integrated into the three-layer pluggable model, which is roughly layered as follows:

- **Scheduling**: Located in the L1 kernel layer, this includes task scheduling and task arrangement. It provides support for upper-layer functions such as elastic scaling, online encryption and desensitization, and MGR detection, and will support more functions in the future.

- **Data ingestion**: Located in the L1 kernel layer, this includes stock data extraction and incremental data acquisition. It supports upper-layer functions such as elastic scaling and online encryption and desensitization, with more features to be supported in the future.

- **The core process of the data pipeline**: Located in the L1 kernel layer, this includes data pipeline metadata and reusable basic components in each step. It can be flexibly configured and assembled to support upper-layer functions such as elastic scalability and online encryption and desensitization. More features will be supported in the future.

- **Elastic scaling, online encryption, and desensitization**: Located in the L2 function layer, it reuses the L1 kernel layer and achieves lightweight functions through configuration and assembly. The SPI interface of some L1 kernel layers is realized through the dependency inversion principle.

- **Implementation of database dialect**: Located in the L3 ecological layer, it includes source end database permission checks, incremental data acquisition, data consistency verification, and SQL statement assembly.

- **Data source abstraction and encapsulation**: These are located in the L1 kernel layer and the L2 function layer. The basic classes and interfaces are located in the L1 kernel layer, while the implementation based on the dependency inversion principle is located in the L2 function layer.

> **Note**
> Elastic scaling is not supported in the following events:
> - If a database table does not have a primary key
> - If the primary key of a database table is a composite primary key
> - If there is no new database cluster at the target end

You are now an elastic scaling master in the making! You're only missing a playing field – that is, how and which real-world issues can you solve with elastic scaling? Read on to learn more.

How to leverage this technology to solve real-world issues

In the preceding sections, we talked about how increased pressure on systems may cause databases to become bottlenecks within an organization. As a solution, we have proposed elastic scaling, a technology that has gained widespread approval. What would be the typical application scenarios for this technology? Or, better yet, would you like to receive some real-world examples where this technology can be applied?

If that's the case, then please read on.

Typical application scenario 1

If an application system is using a traditional single database, with an amount of single-table data that has reached 100 million and is still growing rapidly, the single database continues to be under a high level of pressure, becoming a system bottleneck. Once the database becomes a bottleneck, expanding the application server becomes invalid, and the database needs to be expanded.

In this case, Apache ShardingSphere Proxy can provide help through the following general process:

1. Adjust the proxy configuration so that the existing single database becomes the storage layer of the proxy.
2. The application system will connect to the proxy and use the proxy as a database.
3. Prepare a new database cluster, and install and start the database instance.
4. Adjust the sharding count according to the hardware capability of the new database cluster to trigger the elastic scaling of Apache ShardingSphere.
5. Perform complete database cluster switching through Apache ShardingSphere's elastic scaling function.

Typical application scenario 2

If we expand the database according to scenario 1, but the number of users and visits has increased three times and continues to grow rapidly, the database load continues to be high, creating a system bottleneck, and we must continue to expand the database.

In this case, Apache ShardingSphere Proxy can help. The process is similar to scenario 1:

1. According to the historical data growth rate, plan how many database nodes are required in the next period (such as 1 year), prepare hardware resources, and install and start the database instance.

2. The following steps are the same as *steps 4* and *5* of scenario 1.

Typical application scenario 3

If an application system contains some sensitive data saved in plaintext, to protect the user's sensitive data, even in the case of data leakage, it is necessary to encrypt the data.

In this case, Apache ShardingSphere Proxy can help encrypt and save the stock of sensitive data and subsequent new, sensitive data. It can be completed through the following general process:

1. Similar to scenario 1, first, configure and run the proxy.

2. Add the encrypted rule through DistSQL and configure the encryption option to trigger the Apache ShardingSphere encryption function.

3. Finish encrypting stock data and incremental data through Apache ShardingSphere's online encryption function.

With the last application scenario fully grasped, you now understand elastic scaling and Apache ShardingSphere's implementation of this technology. Now, you are ready to either apply your elastic scaling knowledge to your data environment. Alternatively, you can continue reading and learn more about ShardingSphere's features.

If you have chosen the second option, then please read on to learn more about r ead/write splitting.

Read/write splitting

With the growth of business volume, many applications will encounter the bottleneck of database throughput. It is difficult for a single database to carry a large number of concurrent queries and modifications. At the time of writing, a database cluster with primary-secondary configuration has become an effective scheme. Primary-secondary configuration means that the primary database is responsible for transactional operations such as data writing, modification, and deletion, while the secondary database is responsible for the database architecture of query operations.

The database with the primary-secondary configuration can limit the row lock that's brought by write operations to the primary database and support a large number of queries through the secondary database, greatly improving the performance of the application. Additionally, the multi-primary and multi-secondary database configuration can be adopted to ensure that the system is still available, even if the data node is down and if the database is physically damaged.

However, while the primary-secondary configuration brings the advantages of high reliability and high throughput, it also brings many problems.

The first is the inconsistency of the primary-secondary database data. Because the primary database and the secondary database are asynchronous, there must be a delay in database synchronization, which easily causes data inconsistency.

The second is the complexity brought by database clusters. Database operators, maintenance personnel, and application developers are facing increasingly complex systems. They must consider the configuration of the primary-secondary database to complete business tasks or maintain the database. The following diagram shows the complex topology between the application system and the database when the application system uses data sharding and read/write separation at the same time:

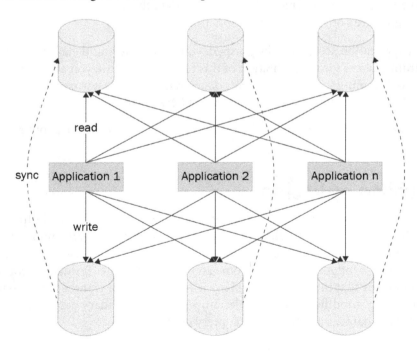

Figure 3.10 – Read/write splitting between databases

Here, we can see that the database cluster is very complex. To solve these complex problems, Apache ShardingSphere provides the read/write splitting function, allowing users to use the database cluster with primary-secondary configuration just as if they were using the native database.

Read/write splitting definition

Read/write splitting refers to routing the user's query operations and transactional write operations to different database nodes to improve the performance of the database system. Read/write splitting supports using a load balancing strategy, which can distribute requests evenly to different database nodes.

In the following sections, we will learn about the key points of the read/write splitting function, see how it works, and check out a few scenarios where this module has been applied.

Key points regarding the read/write splitting function

The read/write splitting function of Apache ShardingSphere includes many database-related concepts, including primary database, secondary database, primary-secondary synchronization, and load balancing strategy.

- **Primary database** refers to the database that's used for adding, updating, and deleting data. At the time of writing, only a single primary database is supported.

- **Secondary database** refers to the database that's used for query data operation, which can support multiple secondary databases.

- **Primary-secondary synchronization** refers to asynchronously synchronizing data from the primary database to the secondary database. Due to the asynchrony of primary-secondary database synchronization, the data of the secondary database and primary database will be inconsistent for a short time.

- A **load balancing strategy** is used to divert query requests to different secondary databases.

How it works

The Apache ShardingSphere read/write splitting module parses the SQL input from the user through the SQL parsing engine and extracts the parsing context. The SQL routing engine identifies whether the current statement is a query statement according to the parsing context, to automatically route the SQL statement to the primary database or secondary database. The SQL execution engine will execute the SQL statement according to the routing results and is responsible for sending SQL to the corresponding data node for efficient execution. ShardingSphere's read/write splitting module has a variety of load balancing algorithms built into it, so that requests can be evenly distributed to each database node, effectively improving the performance and availability of the system.

The following diagram shows the different routing results of two different semantic SQLs. Here, we can see that read operations and write operations are routed to the secondary database and the primary database, respectively. For transactions containing both read and write operations, the ShardingSphere read/write splitting module will also route the corresponding SQL to the primary database to avoid data inconsistency:

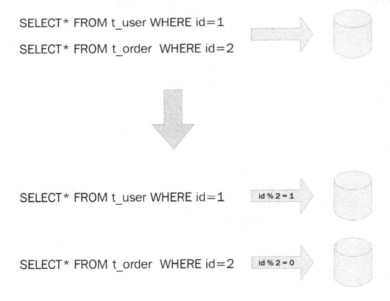

SELECT* FROM t_user WHERE id=1

SELECT* FROM t_order WHERE id=2

SELECT* FROM t_user WHERE id=1   id % 2 = 1

SELECT* FROM t_order WHERE id=2   id % 2 = 0

Figure 3.11 – Semantic SQL routing results

Application scenarios

With the exponential growth of data volume, the linear growth brought by Moore's law to chip performance has made it increasingly difficult to meet users' needs in terms of performance, especially in the internet and financial industries.

To solve this problem, architecture adjustment has become the go-to solution. The read/write splitting architecture is favored by many companies because of its high availability and high performance.

To mitigate the cost increase caused by read/write splitting, Apache ShardingSphere can effectively reduce the complexity of read/write splitting, and let users use the database cluster with a primary-secondary configuration, similar to using a database.

Its built-in load balancing strategy can balance the load of each secondary database, to further improve system performance. In addition, the read/write splitting function of Apache ShardingSphere can be used in combination with its sharding function to further reduce the load and take the performance of the whole system to a new level.

The application scenarios for read/write splitting are truly numerous and understanding this feature will prove to be very useful to you.

Summary

In this chapter, you learned about the pluggable architecture of Apache ShardingSphere and started exploring the features that make up this architecture. By exploring features such as distributed databases, data sharding, SQL optimization, and elastic scaling, you now understand the component-based nature of Apache ShardingSphere.

This project was built to provide you with any feature you may need, without imposing features on you that you may not need, thus giving you the power to build a ShardingSphere. You are now gaining the necessary knowledge to make informed decisions about which features to include in your version.

Since this chapter focused on the main features that will be necessary to build your distributed database and perform data sharding, in the next chapter, we will introduce the remaining major features that make up Apache ShardingSphere.

4

Key Features and Use Cases – Focusing on Performance and Security

The chapter you are about to undertake will further enhance your understanding of the ShardingSphere ecosystem and empower your decision-making to assemble your ideal database solution.

We are going to introduce you to the remaining major features of ShardingSphere, namely high availability, data encryption, and observability.

ShardingSphere is built with flexibility in mind, and the desire to empower you to build a solution that best works for you without imposing undesired bloatware on you. Our community has developed these features for those looking to improve and monitor performance as well as enhance their system's security.

In this chapter, we will cover the following topics:

- Understanding High Availability
- Introducing data encryption and decryption
- User authentication
- SQL authority
- Database and app online tracing
- Database gateway
- Distributed SQL
- Understanding the cluster mode
- Cluster management
- Observability

> **You can find the complete code file here:**
> `https://github.com/PacktPublishing/A-Definitive-Guide-to-Apache-ShardingSphere`

Understanding High Availability

High Availability (**HA**) is one of the important factors that must be considered in the architecture design of distributed systems. The HA of distributed systems ensures that the system can provide services stably. How we minimize the time of service unavailability through design is the goal of HA. As a distributed database service solution, Apache ShardingSphere can provide a distributed capability for the underlying database.

In addition to a HA scheme that can be flexibly compatible with each database and real-time perception of the state changes of the underlying database cluster, the HA of its own services also needs to be guaranteed in the deployment of the production environment.

In the following sections, we will introduce you to HA in general, or refresh your memory if you are already somewhat familiar with the concept. Once complete, you will dive into ShardingSphere's HA.

Database HA

The database is the infrastructure of the software system architecture. The HA of the database is the basis to ensure the security and continuity of business data. For any database, there are many HA schemes for you to choose from. Usually, the HA schemes of the database aim to solve the following list of problems:

- In the case of unexpected downtime or an interruption to the database, the downtime can be reduced as much as needed to ensure that the business will not be interrupted due to the failure of the database.

- The data of the primary node and the backup node shall maintain real-time or final consistency.

- When the database role is switched, the data before and after switching should be consistent.

Currently, read/write splitting is also an essential step for internet software in the data architecture layer. When integrating traditional database HA solutions, usually, it is necessary to deal with the scenario of read/write splitting.

ShardingSphere HA

Apache ShardingSphere adopts the architecture mode of separation of computing and storage. The storage layer is the underlying database cluster managed by ShardingSphere, and the computing layer relies on part of the computing power provided by the underlying database cluster to realize powerful incremental functions. Therefore, the HA of ShardingSphere depends on whether the computing layer and storage layer can provide stable services at the same time. In ShardingSphere, the database clusters of the storage layer are called **storage nodes**, while the ShardingSphere-Proxy deployment of the computing layer is called **computing nodes**.

Taking ShardingSphere-Proxy as an example, the following diagram presents the overall topography of a typical system:

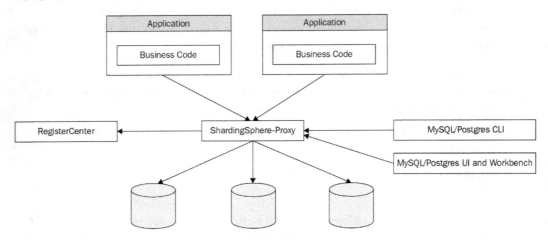

Figure 4.1 – The ShardingSphere-Proxy topography

Thanks to the powerful pluggable architecture of ShardingSphere, the read/write splitting function provided by ShardingSphere can be flexibly combined with HA functions. When the primary-secondary relationship of the underlying database changes, ShardingSphere can automatically route the corresponding SQL to the new database.

With this, in the following sections, let's dive deeper into learning more about computing and storage nodes, how they work, and their application in real-life scenarios.

Computing nodes

As a computing node, ShardingSphere-Proxy is stateless. Therefore, in the same cluster, you can expand the cluster horizontally by adding computing nodes in real time to improve the throughput of the entire ShardingSphere distributed database service. During a low-peak period, users can also release resources by offline computing nodes in real time.

For the HA of the computing node, load balancing software such as *haproxy* or *keepalived* can be deployed on the upper layer of the computing node.

Storage nodes

ShardingSphere's storage node is the underlying database node. The HA of the storage node is guaranteed by the underlying database itself. For MySQL, it can use its own **MySQL Group Replication (MGR)** plugin or third-party orchestrator and other HA schemes, while ShardingSphere can easily integrate the HA schemes of these databases through flexible configuration. Through the internal exploration mechanism, the status information of the underlying database is queried in real time and updated to the ShardingSphere cluster.

How it works

ShardingSphere uses the standard **Serial Peripheral Interface (SPI)** to integrate the HA schemes of various databases. Each scheme needs to implement the interface of database discovery. During initialization, ShardingSphere will use the user-configured HA scheme to create a scheduling task based on ElasticJob, and by default, execute this task every 5 seconds to query the underlying database status, and update the results to the ShardingSphere cluster in real time.

Application scenarios

In real application scenarios, usually, the HA of the ShardingSphere storage nodes is combined with the read/write splitting function to realize the real-time management of the status of primary-secondary nodes. ShardingSphere can automatically handle scenarios such as primary-secondary switching, having the primary database online, and having the secondary database offline. The following configurations show the scenarios of the combination of read/write splitting and HA:

```
rules:
!DB_DISCOVERY
  dataSources:
    pr_ds:
      dataSourceNames:
        - ds_0
        - ds_1
        - ds_2
      discoveryTypeName: mgr
  discoveryTypes:
    mgr:
      type: MGR
      props:
```

```
        groupName: 92504d5b-6dec-11e8-91ea-246e9612aaf1
        zkServerLists: 'localhost:2181'
        keepAliveCron: '0/5 * * * ?'


discoveryTypes:
    mgr:
      type: MGR
      props:
        groupName: 92504d5b-6dec-11e8-91ea-246e9612aaf1
```

The preceding code is the perfect example of a combination of read/write splitting and HA. Now you have acquired the necessary knowledge to move on to more complex features such as encryption.

Introducing data encryption and decryption

Information technology has become increasingly widespread and advanced, making more and more corporations aware of the value of their data assets and the importance of data security.

The need for data encryption and decryption has naturally become commonplace in various enterprises, and data protection methods where you would encode information with an encryption key can allow you to safely manage data.

Considering the current industry demands and pain points, Apache ShardingSphere provides an integrated plan for data encryption. The low-cost, transparent, and safe solution can help all businesses meet their demands for data encryption and decryption.

In the next sections, first, you will gain an understanding of what data encryption is, its application scenarios in both new and established businesses, and finally, move on to ShardingSphere's encryption feature and its workflow.

What are data encryption and decryption?

Data encryption is based on encryption rules and algorithms to transform data and protect user data privacy, while data decryption uses decryption rules to decode encrypted data. Apache ShardingSphere's encryption and decryption module can get the fields that you want to encrypt from SQL, call the encryption algorithm to encrypt the fields, and then store them in a database.

Later, when the user sends a query, ShardingSphere can call the decryption algorithm, decrypt the encrypted column, and return the plaintext data to the user. During the whole process, users do not need to perceive encryption and decryption. Additionally, the module provides complete solutions for both new and established businesses.

In practice, if we were to clarify the differences between new and established businesses, we would find two business scenarios for data encryption. The first business scenario reveals the difficulty of upgrading an old encryption method. Now imagine that Company A is a new business. Everything is new in Company A. If it requires data encryption, its developer team will probably choose a simple data encryption solution just to meet the basic encryption requirements. However, Company A grows rapidly, so the original encryption scheme is not good enough to meet the requirements found in its new business scenarios. Now, it's time for Company A to transform the business system on a large scale, but the cost is too high.

The other scenario describes how hard it will be for a piece of software to be available for use without data encryption to add the data encryption feature into its mature business system. In the past, some businesses chose to store data in plain text, but they now need data encryption. To add data encryption, they have to solve a series of problems such as old data migration encryption (or data cleaning) and recreating related SQL. It's a really complex process, not to mention zero downtime. To support zero-downtime upgrades of a key business, programmers have to establish a pre-release environment and prepare a rollback plan, leading to skyrocketing costs.

Key components

To understand Apache ShardingSphere's encryption and decryption feature, we expect you to understand the related rules and configurations, as shown in the following diagram:

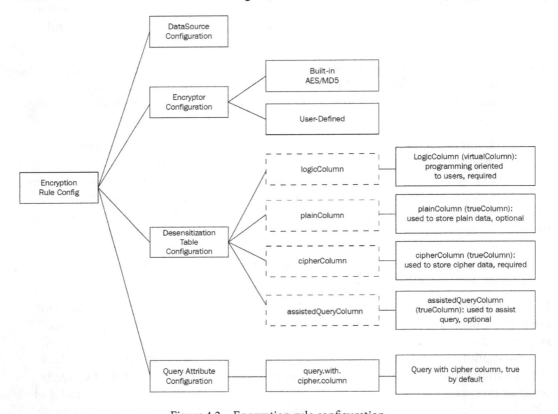

Figure 4.2 – Encryption rule configuration

Let's dig deeper and get an understanding of the terms presented in the preceding diagram:

- **Data Source Configuration**: This is used to configure the data source.

- **Encryption Algorithm Configuration**: This refers to the chosen encryption algorithm. Now, ShardingSphere has two built-in encryption algorithms, that is, **advanced encryption standard** (**AES**) and **message-digest algorithm** (**MD5**). Of course, when necessary, users can leverage the API to develop a custom encryption algorithm.

- **Encryption Table Configuration**: This is used to show the following:

  - Which column is the cipher column storing encrypted data

- Which column is the plain column storing unencrypted data
- Which column is the logic column where the user writes SQL statements

- **Logic Column**: This is the logical name used to calculate an encryption/decryption column. It's also the logical identifier of the column in SQL. Logical columns include cipher columns (required), assisted query columns (optional), and plaintext columns (optional).

- **Cipher Column**: This is the encrypted column.

- **Assisted Query Column**: Literally, the assisted query column is designed to facilitate queries. In terms of some non-idempotent encryption algorithms with higher security levels, ShardingSphere also provides queries with an irreversible idempotent column.

- **Plain Column**: This is the type of column that stores plaintext. It can still provide services during an encrypted data migration process. The user can delete it when data cleaning ends.

Now we can proceed to understand what implementing encryption with Apache ShardingSphere will look like.

Workflow

In Apache ShardingSphere, the workflow of the encryption and decryption feature is shown in the following diagram:

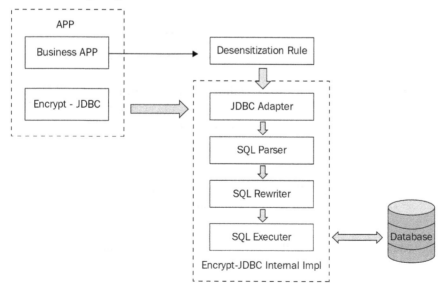

Figure 4.3 – The encryption workflow

At first, the **SQL Parser Engine** parses your SQL input and extracts the contextual information. Then, **SQL Rewriter Engine** rewrites the SQL according to the contextual information and the configuration rules of encryption and decryption. During the process, it rewrites logical columns and encrypts the plaintext. Then, the results are executed efficiently and securely by **SQL Executor Engine**. As you can see, the entire encryption and decryption process is completely transparent to users and perfect for safe data storage:

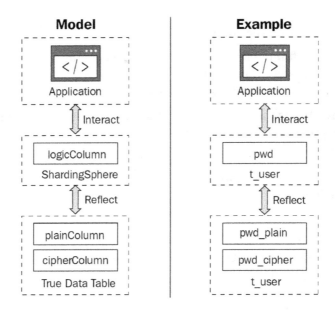

Figure 4.4 – The encryption and decryption process

The workflow process of encryption and decryption is showcased in the following example. Now, let's assume that there is a t_user table in your database. This table contains two fields: pwd_plain (to store plaintext data, that is, the fields to be encrypted) and pwd_cipher (to store ciphertext data, that is, the encrypted fields). At the same time, it's required to define logicColumn as pwd. As the user, you should write the following SQL command: INSERT INTO t_user SET pwd = '123'.

When Apache ShardingSphere receives the SQL command, it will complete the conversion process (as shown in *Figure 4.4*) to rewrite the actual SQL. From *Figure 4.4*, you can understand how addition, deletion, modification, and querying work in the encryption module.

Figure 4.5 illustrates the processing flows and logic of the conversion:

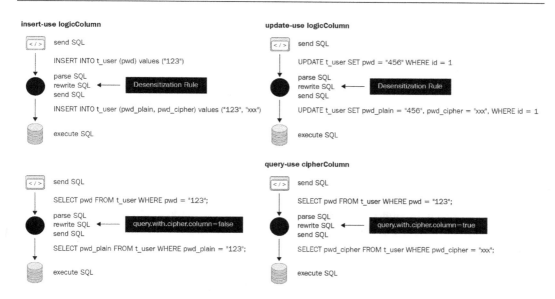

Figure 4.5 – Conversion flow and logic

Let's look at some application scenarios where you would apply the encryption workflow.

Application scenarios

Apache ShardingSphere's encryption and decryption module has already provided users with an integrated solution for encryption and decryption. Both newly launched products and mature products can easily access the encryption and decryption module.

Given that it's not necessary for newly-released products to adopt data migration and data transformation, users only need to access the encryption and decryption module and create relevant configurations and rules. After you select the right encryption algorithm (for example, AES), you just need to configure logical columns (which are used to write user-oriented SQL) and cipher columns (which are used to store encrypted data in data tables). The logical column and the cipher column don't have to be the same. The configuration is shown as follows:

```
- !ENCRYPT
  encryptors:
    aes_encryptor:
```

```
      type: AES
      props:
        aes-key-value: 123456abc
   tables:
     t_user:
       columns:
         pwd:
           cipherColumn: pwd
           encryptorName: aes_encryptor
```

When you access Apache ShardingSphere's encryption feature, the encrypted pwd field will be automatically stored in the database. When you send a query, ShardingSphere will automatically help you to decrypt it and obtain the plaintext. Later, if a user wants to change the encryption algorithm, they only need to modify some related configuration.

So, how does the encryption feature help mature businesses? The answer is that ShardingSphere adds a plaintext column configuration and a cipher column query configuration to smooth the entire encryption transformation process and minimize the costs. Apache ShardingSphere's encryption plan for JD Technology is a good case in point. We'd like to showcase how ShardingSphere solves the encryption transformation issues of JD Technology's mature system.

In this case, the table to be encrypted is t_user, and the field to be encrypted is pwd. The unencrypted field is now stored in the database. At first, we need to replace the standard JDBC with Apache ShardingSphere JDBC. Since Apache ShardingSphere JDBC per se is a standard JDBC API, such change almost costs nothing. Then, we need to add the pwd_cipher field (that is, the field we want to encrypt) to the original table and complete the configuration, as shown in the following code block:

```
-!ENCRYPT
  encryptors:
   aes_encryptor:
     type: AES
     props:
       aes-key-value: 123456abc
  tables:
   t_user:
     columns:
       pwd:
         plainColumn: pwd
```

```
        cipherColumn: pwd_cipher
        encryptorName: aes_encryptor
        queryWithCipherColumn: false
```

As shown in the preceding configuration, we set `logicColumn` as pwd so that users don't need to change the relevant SQL statements. Next, we restart the configuration. At the stage of data insertion, the unencrypted data is still stored in the pwd field, while the encrypted field is stored in the `pwd_cipher` field. In addition, `queryWithCipherColumn` is configured as `False`. Thus, the unencrypted pwd field can still be used for queries and other operations. The next step is encrypted migration of old data, that is, **data cleaning**. However, Apache ShardingSphere has not provided a tool for migration and data cleaning yet, so users still have to process old data by themselves.

After data migration, users can set `queryWithCipherColumn` as `True`, and perform a query or another operation on the encrypted data. At this point, we have completed the majority of the transformation and formed an encrypted dataflow system where plaintext is a backup for rollback. Even if a system failure occurs later, we only need to set `queryWithCipherColumn` as `False` again to form a plaintext dataflow system.

After the system runs stably for a period of time, we can remove the plaintext configuration and delete the unencrypted fields in the database to complete the encryption transformation. The final configuration is as follows:

```
- !ENCRYPT
  encryptors:
   aes_encryptor:
     type: AES
     props:
       aes-key-value: 123456abc
  tables:
   t_user:
     columns:
       pwd:
         cipherColumn: pwd_cipher
         encryptorName: aes_encryptor
```

In Apache ShardingSphere, the encryption and decryption module provides you with an automated and transparent encryption method in order to free you from having to go through the encryption implementation details. With Apache ShardingSphere, you can easily use its built-in encryption and decryption algorithms to meet your requirements for data encryption. Of course, you can also develop your own encryption algorithms via an API. Impressively, both new systems and mature systems can easily access the encryption and decryption module.

In addition, Apache ShardingSphere's encryption module provides users with an enhanced encryption algorithm that is more complex and safer than regular encryption algorithms. When you use the enhanced algorithm, even two identical pieces of data can be encrypted differently. The principle of the enhanced algorithm is to encrypt/decrypt original data and variables together. A timestamp is one of the variables. However, since identical data has different encryption results, it's impossible for queries with cipher data to return all of the data. To solve this issue, Apache ShardingSphere creates the **Assisted Query Column** concept, which can store the original data in an irreversible encryption process: when a user queries encrypted fields, the **Assisted Query Column** concept can assist the user in a query. Of course, all of this is completely transparent to the user.

As Apache ShardingSphere is transparent, automated, and scalable, it enables you to complete low-cost encryption transformation, and even if security requirements change later, you can quickly and conveniently adjust to the changes.

While remaining on the theme of data safety and protection, you might be wondering about another fundamental aspect – user authentication. The next section will teach you everything there is to know about user authentication and Apache ShardingSphere.

User authentication

Identity authentication is the cornerstone of database security protection: only authenticated users are allowed to operate databases. On the other hand, databases can also check the user's identity to determine whether the current operation has been authorized or not.

As the distributed database succeeds the centralized **Database Management System (DBMS)**, user authentication is facing new challenges.

This section will help you understand the authentication mechanism of distributed databases and related features of Apache ShardingSphere.

Authentication of DBMS versus distributed database

The following section introduces you to user ID storage and the differences in its implementation between a DBMS and a distributed database. You might already be familiar with DBMS user ID storage as it's the most widespread, which makes this distinction even more important for you to understand before moving forward.

First, we will start with the main differences and then move on to the mechanism, configuration, and workflow.

User ID storage

In centralized DBMS, user authentication information is often stored in a special data table. For example, as a famous centralized database by design, MySQL stores user information in the mysql.user system table:

```
mysql> select Host, User from mysql.`user`;
+-----------+----------------+
| Host      | User           |
+-----------+----------------+
| %         | root           |
| localhost | mysql.session  |
| localhost | mysql.sys      |
| localhost | root           |
+-----------+----------------+
4 rows in set (0.00 sec)
```

Accordingly, when the client side establishes a connection to MySQL, the MySQL server will match the request information with the data in the mysql.user table to determine whether the user is allowed to connect. Of course, this table also contains encrypted user password strings because a user cannot be authenticated without their matching password.

However, given database clusters managed by ShardingSphere-Proxy or other distributed systems, the story is totally different.

The number of database resources managed by ShardingSphere-Proxy might be zero or even several thousand. In this situation, if you still insist on using tables in centralized databases such as `mysql.user` to store ID data, you might face the following questions:

- Without resources, where is ID data stored?
- If you have thousands of resources, where is the ID data stored?
- When resources increase or decrease, should the ID data be synchronized?

Now that we have clarified the user ID storage differences between DBMS and a distributed database, we should also consider the differences in their protocols and encryption algorithms. The next section clarifies everything you should consider when it comes to both protocols and encryption algorithms.

Different protocols and encryption algorithms

In addition to the issue of user ID information storage, protocol adaptation is another difficult problem facing distributed databases.

ShardingSphere-Proxy has already supported different database protocols such as MySQL and PostgreSQL, so users can directly connect MySQL's client side or PostgreSQL's client side to ShardingSphere-Proxy.

When different database client sides initiate TCP handshakes to establish connections, they will encrypt user passwords with different rules. To solve this issue, ShardingSphere-Proxy has been developed to recognize different database protocols and use respective encryption algorithms to verify passwords along with user IDs. The following sections introduce the user ID information starting from the mechanism, then moving to the workflow and configuration, and finally, to the application scenarios.

Mechanism

When we talk about mechanisms, we actually mean that there are a few different points that you should get familiar with, in order to understand the mechanism. The following subsections will introduce you to the user ID information storage and the adaptation of different protocols.

Storing your user ID information

Of course, it is not appropriate to store user ID information in one or several database resources. Distributed databases need to provide a centralized ID information storage space. In fact, storing the data within a registry center is a good idea:

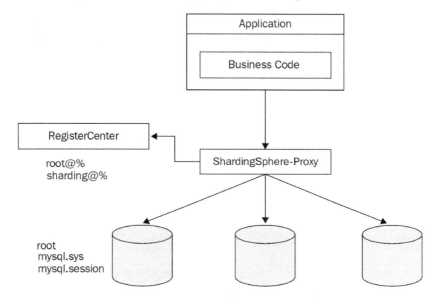

Figure 4.6 – Registry center data storage

As shown in the preceding diagram, when a distributed database system has just been established, each DBMS still has its own user information that is used to establish its connections with ShardingSphere-Proxy. When an application wants to use the distributed database, it uses the user and password data stored in the registry center and connects to ShardingSphere-Proxy. That's how ID authentication in a distributed database becomes real.

Adaptation of different protocols

Apache ShardingSphere is known for its powerful SPI mechanism. The mechanism is beneficial for fixing the protocol adaptation issue: ShardingSphere-Proxy provides different types of authentication engines for different database protocols. Each authentication engine adopts the native implementation method, that is, different encryption algorithms are implemented for each type of database to ensure the best customer experience after database migration. *Figure 4.7* gives you an overview of the compatibility of ShardingSphere with different protocols:

Figure 4.7 – ShardingSphere protocol adaptability

The previous screenshot illustrates the protocols that Apache ShardingSphere is compatible with, and this gives you a complete overview of the mechanisms that make up the user authentication feature. Now, let's move on to the workflow and application scenarios of this feature.

Workflow

Apache ShardingSphere abstracts rules from many capabilities and loads the rules in the form of SPIs on the kernel. Such design largely extends custom capabilities.

The global authority rule is an indispensable rule for ShardingSphere-Proxy booting. Currently, ShardingSphere provides users with a default `AuthorityRule` implementation.

Now, let's jump into the configuration aspect so that you can set up your user ID storage with your ShardingSphere ecosystem.

Configuration

First, the user needs to set up the initial user in `server.yaml` under the working directory of ShardingSphere-Proxy:

```
rules:
  - !AUTHORITY
    users:
      - root@%:root
      - sharding@:sharding
    provider:
      type: ALL_PRIVILEGES_PERMITTED
```

In the preceding configuration, you can see that Proxy initializes two users, `root@%` and `sharding@`. Their respective initial passwords are `root` and `sharding`.

Note that `%` in `root@%` means that ShardingSphere-Proxy can be connected to via any host address. When nothing follows @, the default value is `%`. You can also use `user1@ localhost` to restrict `user1` to local logins only.

Initialization

When ShardingSphere-Proxy boots, first, it persists local configuration information to the registry center (if the `Overwrite` value of `Mode` is set as `True`). After that, the required configuration information is obtained from the registry center to construct ShardingSphere-Proxy's metadata, including the user ID information that we just configured.

According to the provider configuration, ShardingSphere uses the specified authority provider to build user metadata, including user ID and authorization information stored in memory in the form of a map. Therefore, when a user requests to connect or to obtain authorization, the system can quickly respond to their request, that is, there's no second query via the registry center.

Implementing ID authorization

One of the possible requirements you might have is to implement ID authorization. This section will guide you through Apache ShardingSphere's ID authorization.

In terms of MySQL client-side logins, ShardingSphere performs the following operations for the user:

1. Uses the username entered by the user to search whether the user exists in metadata; if not, refuses the connection.

2. Compares the host address that the user logs in to determine whether it complies with user configuration; if not, rejects the connection.

3. Based on the encryption algorithm, it encrypts the user password in the metadata. Then, it compares the encrypted password with the received one. If the two are inconsistent, it rejects the connection.

4. If the user specifies a schema, it queries the user authorization list to check whether they have the permission to connect to the schema; if not, it refuses the connection.

5. After the preceding processes have been completed, the user can finally log in and connect to the specified logical schema.

In this section, we discussed why Apache ShardingSphere's user ID authentication mechanism is the first step toward achieving database security protection. By configuring different users and restricting different login addresses, ShardingSphere can provide enterprises with varied data security degrees.

In the future, with the help of its active community, Apache ShardingSphere is expected to have more security mechanisms, such as configuration encryption and custom encryption algorithms. Of course, the flexible SPI mechanism of ShardingSphere allows users to develop custom authentication plugins and build their own *access control system* of distributed databases.

In the next section, we will introduce you to the SQL Authority feature and its scalability. This is an essential feature if you are interested in securing your data and distributed database.

SQL Authority

It's understood that by following the evolution from centralized databases to distributed ones, the user ID authentication mechanism has undergone dramatic changes. Similarly, any logic database provided by Apache ShardingSphere does not exist in a certain database resource, so ShardingSphere-Proxy is required to centralize the processing of user permission verification.

In the previous chapter, in terms of user login authentication, we explained that if the user specifies the schema to be connected, ShardingSphere can determine whether the user has the permissions based on the authorization information. How can it do that?

In this chapter, we will showcase Apache ShardingSphere's SQL Authority feature and its scalability.

Defining SQL Authority

We can describe SQL Authority like this: after receiving a user's SQL command, Apache ShardingSphere checks whether the user has authority based on the data type and data scope requested by the command. It then decides to allow or reject the operation.

Mechanism

In the *User authentication* section, we described `AuthorityRule` and `AuthorityProvider`. In fact, in addition to user ID information, user authorization information is also controlled by `AuthorityProvider`.

Currently, aside from the default `ALL_PRIVILEGES_PERMITTED` type, Apache ShardingSphere also provides the `AuthorityProvider` `SCHEMA_PRIVILEGES_PERMITTED` type to control schema authorization.

To use `SCHEMA_PRIVILEGES_PERMITTED`, the configuration method is as follows:

```
rules:
  - !AUTHORITY
    users:
      - root@:root
      - user1@:user1
      - user1@127.0.0.1:user1
    provider:
      type: SCHEMA_PRIVILEGES_PERMITTED
      props:
        user-schema-mappings: root@=test, user1@127.0.0.1=db_
dal_admin, user1@=test
```

The configuration contains the following key points:

- When the user `root` connects to ShardingSphere from any host, they have permission to access the schema named `test`.

- When `user1` connects to ShardingSphere from `127.0.0.1`, they have the authority to access the schema named `db_dal_admin`.

- When `user1` connects to ShardingSphere from any host, they have the right to use the schema named `test`.

When other unauthorized situations occur, any connection will be refused, such as the following:

- `show databases`

- `use database`

- `select * from database.table`

Apart from the login scenario, a schema permission check is also performed when a user enters the following SQL statements:

- `show databases`

- `use database`

- `select * from database.table`

When the SQL Authority engine finds that the SQL statement input requests the database resource, it will use the interface provided by `AuthorityProvider` to perform permission checks to protect user data security from various angles.

Planned development

Since `AuthorityProvider` is one of the ShardingSphere SPIs, its future development is given a lot of room with multiple possible avenues.

Besides database-level authorization control, the community might develop fine-grained authorization methods such as table-level authorization and column-level authorization.

In the future, `AuthorityRule` will be linked with DistSQL, a new feature just released in version 5.0.0 to explore more flexible user and authorization management methods and make it even more convenient for users.

Application scenarios

User authentication plus SQL Authority contributes to Apache ShardingSphere's integrated database-level security solution.

Concurrently, ShardingSphere abstracts complete top-level interfaces that enable the community and users to implement various levels of data security control and provide technical support for SQL auditing and other application security management scenarios such as password protection and data encryption.

You now have acquired a foundational understanding of data encryption and how Apache ShardingSphere implements data encryption and decryption. You may or may not require this feature, and keeping in mind that ShardingSphere is built on a completely pluggable architecture, you may or may not include it in your version of Apache ShardingSphere.

Database and app online tracing

The distributed tracing system is designed based on the Google Dapper paper. There are already many relatively mature applications of the system, such as Zipkin by Twitter, SkyWalking by Apache, and CAT by Meituan-Dianping.

The following sections will introduce you to how a database and app online tracing work, and then show you how Apache ShardingSphere implements this feature.

How it works

A distributed scheduling chain turns one distributed request into multiple scheduling chains. In the scheduling of one distributed request, such as time consumption on each node, the machine receives the request and the status of the request on each service node can be seen in the backend.

The following is a diagram of one distributed request scheduling chain quoted in Google's paper *Dapper, A Large-Scale Distributed Systems Tracing Infrastructure*:

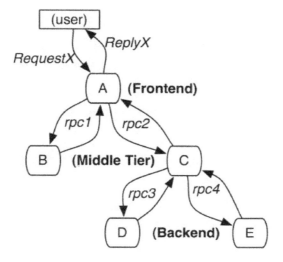

Figure 4.8 – A distributed request scheduling chain

> **Citation**
>
> Dapper, a Large-Scale Distributed Systems Tracing Infrastructure, Benjamin H. Sigelman and Luiz André Barroso and Mike Burrows and Pat Stephenson and Manoj Plakal and Donald Beaver and Saul Jaspan and Chandan Shanbhag, 2010, `https://research.google.com/archive/papers/dapper-2010-1.pdf`, Google, Inc.

As you can see in *Figure 4.8*, a request made by a user takes a path that goes through various nodes, going all the way from the frontend to the backend and back. Monitoring all of these nodes means having total monitoring over your system. In the next sections, we will introduce you to the monitoring solution that you can integrate into your ShardingSphere solution.

A total synthetic monitoring solution

This section will introduce you to **synthetic monitoring solution**, another feature of ShardingSphere's pluggable architecture. You will learn what we mean and how we implement database monitoring, tracing, and the shadow database.

The feature includes three layers for a comprehensive solution. The following list introduces you to each layer and its function:

- **Gateway layer** (cyborg-flow-gateway): This is achieved by Apache APISIX. It is able to tag data and distribute and pass tags to the context of link scheduling.
- **Tracing service** (cyborg-agent, cyborg-dashboard): This is achieved by Apache SkyWalking and can pass stress testing tags throughout the link.
- **Shadow database** (cyborg-database-shadow): This is achieved by Apache ShardingSphere and can isolate data based on stress testing tags.

When a stress testing request is made to the gateway layer (cyborg-flow-gateway), tags will be made to stress testing data according to the configuration within the configuration file. The tags will be passed to distributed link system and flow information will be seen (cyborg-dashboard) throughout the whole link. Before requesting the database, the link tracing service will add stress testing tags to execute SQL through HINT:

```
INSERT INTO table_name (column1,...) VALUES (value1...)
/*foo:bar,...*/;
```

When requests are made to the database proxy layer (cyborg-database-shadow), they will be routed to the corresponding shadow databases according to stress testing tags.

Now that you understand ShardingSphere's interpretation of database monitoring, we can proceed to understand the database gateway.

Database gateway

Unified operation of backend heterogeneous databases through the database gateway can minimize the impact of a fragmented database.

In addition to high performance, fusing, rate limit, blacklisting, and whitelisting, among other capabilities, the database gateway should also have capabilities such as unified access, transparent data access, SQL dialect conversion, and routing decisions. Furthermore, the quintessence of database proxy sustainability lies in an open ecosystem and flexible extensibility.

In this section, we will follow our pattern by introducing the mechanism, to then dive deeper into the components. However, this time, first, we should start with a little background. Considering the significant transformations that databases have been undergoing recently, we need to gain an understanding of what a database gateway is. The next section allows you to do just that.

Understanding the database gateway

Currently, as the distributed database is in its transformation period, although unified products remain the goal pursued by database providers, fragmented databases are becoming the trend. Leveraging their advantages in their respective fields is key for database products to gain a foothold in the industry.

Growing the diversity of the database brings new demands. Neutral gateway products can be built on top of databases to effectively reduce use costs brought by the differences between heterogeneous databases, ultimately allowing you to fully leverage diverse databases' strengths.

Gateway is not a new product. It should have standard capabilities including fusing, rate limits, blacklisting, and whitelisting. Considering their functions of traffic bearing and traffic allocation, gateway products should have non-functional features such as stability, high performance, and security. The database gateway further enhances database functions based on an existing gateway. The biggest difference between a database gateway and a traffic gateway is the diversity of access protocols and the statefulness of target nodes.

A deep dive into the database gateway mechanism

Apache ShardingSphere provides the basic functions of the database gateway mechanism, and other functions that are required to be a digital gateway will be completed in its future version. The database gateway includes major functions such as support for heterogeneous databases, database storage adaptor docking, SQL dialect conversion, routing decisions, and flexible extension. In this section, the major functions and realization mechanisms of the database gateway and the relative route planning of Apache ShardingSphere will be briefly discussed.

Support for heterogeneous databases

Apache ShardingSphere now supports multiple data protocols and SQL dialects.

Engineers can access Apache ShardingSphere through MySQL and PostgreSQL. Additionally, other databases that support similar protocols, such as **MariaDB** and **TiDB**, which support the MySQL protocol, and **openGauss** and **CockroachDB**, which support the PostgreSQL protocol, can also be directly accessed. Apache ShardingSphere simulates the target database through the realization of the binary interactive protocol of the database.

Engineers can use MySQL, PostgreSQL, Oracle, SQL Server, and SQL dialects that meet SQL92 standards to access Apache ShardingSphere. The SQL grammar file is defined by ANTLR, generated through code, and translates SQL into an abstract syntax tree and visitor model through the classic Lexer + Parser plan.

Database storage adaptor docking

Currently, Apache ShardingSphere supports access to the backend database through JDBC. JDBC is a standard interface that allows Java to access databases, supports multiple databases, and can support databases using JDBC protocol without modifying any code in theory.

In practice, although JDBC has a unified interface at the code level, the SQL used to access the databases hasn't been unified. Therefore, in addition to supporting backend databases that support SQL, databases with SQL that are not supported by Apache ShardingSphere can be accessed through SQL92 standards. In other words, except for MySQL, PostgreSQL, Oracle, SQL Server, and databases that support similar dialects that are supported by Apache ShardingSphere, other databases could encounter access errors if they do not enable their dialects to support SQL92 standards. Additionally, if other databases support extra functions in addition to the SQL92 dialect, they will not be recognized by Apache ShardingSphere. Based on meeting the basic requirements of backend databases, engineers can easily access these backend databases through the required SQL dialects.

Although most relational databases and some NoSQL databases support access through JDBC, there are still some databases that do not support access through JDBC. In Apache ShardingSphere's route planning, access models that are targeted at specific databases will be supported, which strive to provide diverse support for databases.

SQL dialect conversion

While accessing a specific database through certain database protocols and SQL dialects could achieve the basic functions of a database protocol, it does not remove the barriers between databases. In Apache ShardingSphere's route planning, the SQL dialect conversion function will be provided to remove the barriers between heterogeneous databases.

For example, engineers could access the data in the HBase cluster through the MySQL protocol and SQL dialect so that they can truly achieve HTAP. Additionally, they could visit other relational databases through the PostgreSQL protocol and SQL dialect, realizing low-cost database migration. In practice, Apache ShardingSphere regenerates the SQL by matching the parsed SQL **Abstract Syntax Tree** (**AST**) with the conversion rules of other database SQL dialects.

Indeed, SQL conversion cannot entirely achieve dialect conversion among all databases. Operators and functions exclusive to a certain database can only be thrown an `UnsupportedOperationException` exception. Nevertheless, the ability of dialect conversion can make correspondence and collaboration between heterogenous databases more convenient. This way, the database gateway is no longer oriented to a single database *statically*. Instead, it *dynamically* chooses a suitable database protocol parser and SQL parsing engine according to the database protocol and SQL dialect, making it a truly unified access of hybrid deployment for heterogeneous databases.

Routing decision

Dynamically unified access allows the database gateway to parse requests to suitable databases more intelligently by understanding the SQL syntax, thereby empowering heterogeneous database clusters with hybrid deployments to enjoy stronger computing capability.

Routing decisions are mainly decided by two factors, namely where the data is and which database can better address the existing request.

By matching our metadata information with SQL AST, the heterogeneous database cluster where the data copy is stored can be identified. When data is stored in multiple heterogeneous databases, the database used to address the query request can be decided by recognizing the SQL features and matching the attribute tags of the heterogeneous database; for example, routing aggregations and grouping operations to OLAP-natured databases or routing transaction queries based on the primary key to OLTP-natured databases. Making decisions via rules or query costs is a relatively effective routing decision strategy.

Routing decision capability is the reverse function of SQL dialect conversion capability. Currently, it is not available in Apache ShardingSphere. Apache ShardingSphere has not taken into consideration the capability of its route planning. Therefore, further development plans will be considered after the SQL dialect conversion function has been completed.

Extensibility

Different from the traffic gateway, open and fast docking with diverse databases is an important function of the database gateway. The traffic gateway can route requests to the backend service by identifying a protocol and allocating weights. The biggest difference between the database gateway and stateless service nodes is that the statefulness of the database in the database gateway requires routing requests with precision. In addition, in terms of extensibility, the traffic gateway does not need to understand the contents it transmits or attend to the backend resource operations. In comparison, the database gateway needs to match the detailed operations of different databases.

In terms of what's been done so far, Apache ShardingSphere's extensibility is represented by the database protocol and the extensible SQL dialect. On the to-do list, Apache ShardingSphere's extensibility is represented by the database operation and SQL dialect conversion; in terms of what has not been planned, the routing decision strategy can be determined and developed by engineers.

Developers can achieve docking with new databases and routing decisions through SPI without changing the core code. Apache ShardingSphere always embraces an open source ecosystem and diverse heterogeneous databases with a neutral stance.

The overall architecture of the database gateway is as follows:

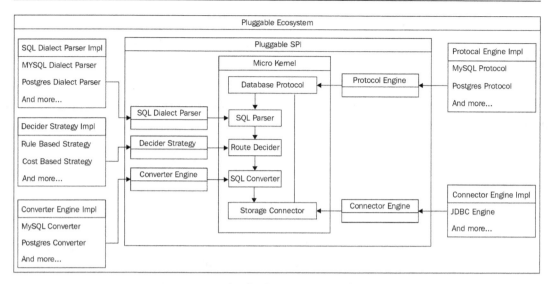

Figure 4.9 – The database gateway architecture

With a database protocol and SQL dialect identification, AST can be routed to the appropriate data source by the routing decision engine, and SQL can be converted into the corresponding data source dialect and executed.

Five top accesses are or will be extracted in Apache ShardingSphere for database protocols, SQL dialect parsing, routing decision strategies (this feature is under development), SQL dialect converters (this feature is under development), and data executor.

Values

The values of the database gateway are unified database access, transparent heterogeneous data access, shielding the database implementation details, and allowing engineers to use all sorts of heterogeneous database clusters such as using a database.

Unified database access

In terms of programming access, with the heterogeneous database protocol, support for SQL dialects and SQL dialect conversion, engineers can switch between different databases by programming in the database gateway. The database gateway is not product-oriented to a single database only. It is neutral, and offers unified access to heterogeneous databases.

Transparent heterogeneous data access

In terms of data storage, engineers do not need to attend to detailed storage space or methods. The database gateway can manage backend databases automatically by matching a powerful routing decision engine with transparent database storage access.

When considering the database gateway, it is important not to confuse the gateway with extensibility. Now you have an understanding of the differences between the two terms and what to expect from the two features.

Distributed SQL

In ShardingSphere 4.x and earlier versions, just like other middleware, people needed to use configuration files to manage ShardingSphere to perform operations such as telling it which logic databases to create, how to split the data, which fields to encrypt, and more. Although most developers are accustomed to using configuration files as a tool to manage middleware, as the entrance to distributed databases, ShardingSphere also serves a large number of operations engineers and DBAs who are more familiar with SQL execution than editing configuration files.

The Database Plus concept is the impetus for Apache ShardingSphere 5.0.0's new interactive language: **Distributed SQL (DistSQL)**.

DistSQL is the built-in language of ShardingSphere that allows users to manage ShardingSphere and all rule configurations through a method that is similar to SQL. This is so that users can operate Apache ShardingSphere in the same way as SQL database operations.

DistSQL contains three types, namely RDL, RQL, and RAL:

- **Resource & Rule Definition Language** (RDL): This is used to create, modify, and delete resources and rules.

- **Resource & Rule Query Language** (RQL): This is used to query resources and rules.

- **Resource & Rule Administration Language** (RAL): This is used to manage features such as hint, transaction type switch, execution plan query, and elastic scaling control.

The following sections will guide you through DistSQL's advantages, along with its implementation in ShardingSphere, and conclude with some application scenarios that'll allow you to see it in action.

Introduction to DistSQL

Without DistSQL, if we want to use ShardingSphere-Proxy for data sharding, we need to configure a YAML file, as follows:

```
# Logic Database Name Used for External Services
schemaName: sharding_db


# Actual Data Source in Use
dataSources:
  ds_0:
    url: jdbc:mysql://127.0.0.1:3306/db0
    username: root
    password:
```

If we take data sharding as an example of the task we want to accomplish with DistSQL, the input code we would use would look like the following:

```
# Specify the Sharding Rule; The example codes means split data
tables in the database into 4 shards
rules:
- !SHARDING
  autoTables:
    t_order:
      actualDataSources: ds_0
      shardingStrategy:
        standard:
          shardingColumn: order_id
          shardingAlgorithmName: t_order_hash_mod
  shardingAlgorithms:
    t_order_hash_mod:
      type: HASH_MOD
      props:
        sharding-count: 4
```

If the YAML configuration file is completed, we can deploy ShardingSphere-Proxy and start using it. If we still need to adjust any resources or rules, we have to update the configuration file and restart Proxy to implement an effective configuration.

With DistSQL, users don't have to configure the files again. They can directly boot ShardingSphere-Proxy and execute the following SQL commands:

```
# Create a Logic Database
CREATE DATABASE sharding_db;
# sharding_db Connect to sharding_db
USE sharding_db;
# Add Data Resource
ADD RESOURCE ds_0 (
    HOST=localhost,
    PORT=3306,
    DB=db0,
    USER=root
);
# Create Sharding Rules
CREATE SHARDING TABLE RULE t_order (
RESOURCES(ds_0),
SHARDING_COLUMN=order_id,TYPE(NAME=hash_mod,
PROPERTIES("sharding-count"=4))
);
```

After we enter the preceding DistSQL command, the rule configuration equivalent to the file operations will be completed. Afterward, you don't need to restart because you can use DistSQL to dynamically add, modify, or delete any resources and rules.

DistSQL considers the habits of database developers and operation engineers and learns from standard SQL's grammar to ensure readability and usability, making DistSQL preserve ShardingSphere features to the greatest extent.

Application scenarios

Distributed SQL is a powerful tool that is also extremely versatile and has a very friendly learning curve. It is applicable to virtually any scenario where normal SQL would be. The following section gives you a few application scenarios to ensure you are familiar with the application scenarios for DistSQL.

First, let's look at *Figure 4.10*:

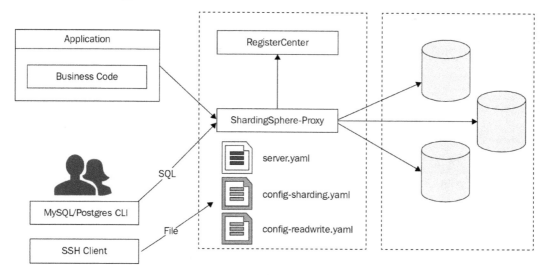

Figure 4.10 – The application system without DistSQL

Without DistSQL, the prevalent condition is detailed as follows:

- An application system is connected to ShardingSphere-Proxy, and data is read and written through the SQL commands.

- Developers or operation engineers connect their SQL client side to ShardingSphere-Proxy and use SQL commands to view data.

- At the same time, they connect the SSH client to the server where the proxy is located in order to perform file-editing operations and configure the resources and rules used by ShardingSphere-Proxy.

- Editing files remotely is not as simple as editing local files.

- Each logic database uses a different configuration file. There might be a lot of files on the server, which makes remembering their locations and differences a challenge.

- As for file updates, it's necessary to restart ShardingSphere-Proxy to make the new configuration effective, but this could interrupt the system.

DistSQL glues ShardingSphere-Proxy onto databases, making them one unit, as shown in the following diagram:

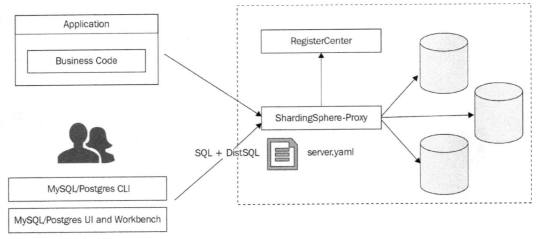

Figure 4.11 – The application system including DistSQL

Once we include DistSQL, the situation changes and we practically eliminate a layer by closing the gap between databases and ShardingSphere-Proxy.

The application system is still connected to ShardingSphere-Proxy, and data is read and written through SQL commands.

Developers or ops engineers no longer need two different tools. They only need to open the SQL client side to view data and manage ShardingSphere at the same time:

- No file editing is required.
- There is no need to worry about the potentially explosive growth of files.

DistSQL can execute real-time update operations; therefore, users don't have to restart ShardingSphere-Proxy and worry about system interruptions.

Additional notes for DistSQL

The previous case where we used data sharding as an example of a task we'd like to accomplish showcased how RDL defines resources and rules. Now, let's explore other impressive features of DistSQL on the basis of these rules:

- **View added resources** (RQL) allows you to query the system and have an immediate overview of the resources included:

```
> SHOW SCHEMA RESOURCES;
```

```
+------+-------+-----------+------+------+----------+
| name | type  | host      | port | db   | attribute |
| ds_0 | MySQL | 127.0.0.1 | 3306 | db0  | ...      |
+------+-------+-----------+------+------+----------+
1 rows in set (0.01 sec)
```

- **View sharding rules** (RQL) allows you to conveniently understand the current sharding rules of the system:

  ```
  > SHOW SHARDING TABLE RULES;
  ```

  ```
  mysql> show sharding table rules;
  +---------+--------------------+--------------------+---
  --------------------+-----------------------------+---
  ----------------------------+
  | table   | actual_data_sources | table_strategy_type |
  table_sharding_column | table_sharding_algorithm_type |
  table_sharding_algorithm_props |
  +---------+--------------------+--------------------+---
  --------------------+-----------------------------+---
  ----------------------------+
  | t_order | ds_0                | hash_mod            |
  order_id              | hash_mod                      |
  sharding-count=4      |
  +---------+--------------------+--------------------+---
  --------------------+-----------------------------+---
  ----------------------------+
  1 row in set (0.01 sec)
  ```

- **Preview distributed execution plans** (RAL) in ShardingSphere can output the execution plans parsed by logical SQL, but it does not actually execute these SQL statements:

  ```
  > PREVIEW select * from t_order;
  ```

  ```
  +-----------------+----------------------------------------
  ----------+
  ```

```
| data_source_name |
sql                                                            |
+------------------+--------------------------------------------
-----------+
| ds_0            | select * from t_order_0 ORDER BY
order_id ASC  |
| ds_0            | select * from t_order_1 ORDER BY
order_id ASC  |
| ds_0            | select * from t_order_2 ORDER BY
order_id ASC  |
| ds_0            | select * from t_order_3 ORDER BY
order_id ASC  |
+------------------+--------------------------------------------
-----------+
4 rows in set (0.01 sec)
```

Aside from these features, DistSQL also supports rule definitions and queries in scenarios such as read/write splitting, data encryption, database discovery, and shadow database stress testing, with the aim to cover all scenarios.

Implications for ShardingSphere

DistSQL was designed with the aim to redefine the boundaries between middleware and a database, which would ultimately allow developers such as yourself to utilize ShardingSphere in the same manner you usually operate your databases.

It was designed with the goal to redefine the boundary between middleware and databases, allowing you and other developers to use Apache ShardingSphere in the same way as database operations. The syntax system of DistSQL acts as a bridge between ShardingSphere and the distributed databases. In the future, when more creative ideas come true, DistSQL is destined to become more powerful and help ShardingSphere become a better database infrastructure.

Understanding cluster mode

Considering user deployment scenarios, Apache ShardingSphere provides three operating modes: **cluster mode**, **memory mode**, and **standalone mode**. Cluster mode is the production deployment method recommended by ShardingSphere. With cluster mode, horizontal scaling is achieved by adding computing nodes, while multi-node deployment is also the basis for high service availability.

In the following sections, you will understand what cluster mode is, as well as its compatibility with other ShardingSphere modes so that you can better integrate it into your pluggable system.

Cluster mode definition

Apart from cluster mode, ShardingSphere also provides its counterpart, that is, its standalone mode deployment. With standalone mode, users can also deploy multiple computing nodes, but unlike with cluster mode, configuration sharding and state coordination cannot be performed among multiple computing nodes in standalone mode. Any configuration or metadata changes only work on the current node, that is, other nodes cannot perceive the modification operations of other computing nodes.

In cluster mode, the feature registry center allows all computing nodes in the cluster to share the same configuration information and metadata information. Therefore, when the shared information changes, the register center can push these changes down to all computing nodes in real time to ensure data consistency in the entire cluster.

Kernel concepts

In the following subsection, we will introduce you to the kernel concept for all the modes, including the registry center, and the mechanism to enable cluster mode.

Operating modes

Cluster mode is one of the ShardingSphere operating modes. The mode is suitable for a production environment deployment. In addition to cluster mode, ShardingSphere also provides memory mode and standalone mode, which are used for integration testing and local development testing, respectively. Unlike standalone mode, memory mode does not persist any metadata and configuration information, in which all modifications take effect in the current thread. The operating modes cover all use scenarios spanning development to testing and production deployment.

The registry center

The registry center is the foundation of the cluster mode implementation. ShardingSphere can share metadata and configurations in cluster mode because it integrates the third-party register components of *Zookeeper* and *Etcd*. Concurrently, it leverages the notification and coordination capabilities of the register center to ensure the cluster synchronization of shared data changes in real time.

The mechanism to enable cluster mode

To enable cluster mode in a production environment, we need to configure the mode tag in server.yaml:

```
mode:
  type: Cluster
  repository:
    type: ZooKeeper
    props:
      namespace: governance_ds
      server-lists: localhost:2181
      retryIntervalMilliseconds: 500
      timeToLiveSeconds: 60
      maxRetries: 3
      operationTimeoutMilliseconds: 500
  overwrite: false
```

Next, to add multiple computing nodes to the cluster, it is necessary to ensure that the configurations of namespace and server-lists are kept the same. Therefore, these computing nodes function in the same cluster.

If users need to use local configuration to initialize or overwrite the configuration in the cluster, they can configure overwrite: true.

Compatibility with other ShardingSphere features

The core capability of cluster mode is to ensure the configuration and data consistency of ShardingSphere in a distributed environment, which is the cornerstone of ShardingSphere's enhanced features in the distributed environment. In terms of DDL, which modifies metadata, after a user executes DDL on a certain computing node, the computing node will update the metadata change to the register center; the coordination function of the register center is used to send the changed message to other computing nodes in the cluster, making the metadata information of all computing nodes in the cluster consistent.

Cluster mode is the foundation for ShardingSphere's production deployment, and it also endows ShardingSphere with distributed capabilities that lay the groundwork for the internet software architecture. Core features such as *horizontal scaling* and *HA* that are provided by cluster mode are also inseparable for distributed systems.

With this knowledge, let's move on to learn how to manage the cluster.

Cluster management

As technologies advance, we not only require big data computing but also 24/7 system services. Accordingly, the old single-node deployment method cannot meet our needs anymore, and the multi-node cluster deployment method is the trend. Additionally, deploying multi-node clusters faces many challenges.

On the one hand, ShardingSphere needs to manage storage nodes, computing nodes, and underlying database nodes in the cluster while it also needs to *refresh*: it detects the latest node changes in real time and adopts the heartbeat detection mechanism to ensure the correctness and availability of the storage, computing, and database nodes. On the other hand, ShardingSphere needs to solve two issues:

- How do you keep consistency among configurations and statuses of different nodes in the cluster?

- How do you guarantee collaborative work between nodes?

ShardingSphere not only integrates the third-party components of **Apache Zookeeper** and **Etcd** but also provides the `ClusterPersistRepository` interface for custom extensions. Additionally, we can use other configuration register components that we like. ShardingSphere leverages the characteristics of Apache Zookeeper and etcd to synchronize different node strategies and rules in the same cluster. Apache Zookeeper and Etcd are used to store configurations of data sources, rules and strategies, and the states of computing nodes and storage nodes, to better manage clusters in ShardingSphere.

Computing nodes and storage nodes are the most important aspects of ShardingSphere. Computing nodes handle switch-on and the fusing of running instances, while storage nodes manage the relationships between primary databases and secondary databases plus the database status (enable or disable). Now, let's see the differences between the two and how to work with them.

Computing nodes

Two subnodes of the `/status/compute_nodes` computing node are listed as follows:

- `/status/compute_nodes/online` stores online running instances.

- `/status/compute_nodes/circuit_breaker` stores breaker running instances.

For both online and breaker running instances, the identity of the running instance is the same: it is composed of the host's IP and port number. When a running instance goes online, it will automatically record its own IP and port number under the `/status/compute_nodes/online` computing node to become a member of the cluster.

Similarly, when the running instance is broken, the instance will also be removed from the `/status/compute_nodes/online` computing node and record its IP and port number in the `/status/compute_nodes/circuit_breaker` computing node.

Storage nodes

The two sub nodes of the `/status/storage_nodes` storage node are listed as follows:

- `/status/storage_nodes/disable` stores the current disabled secondary database.
- `/status/storage_nodes/primary` stores the primary database.

To modify the relationship between the primary database and the secondary database or to disable the secondary database, we can use DistSQL or high availability to automatically sense the primary-secondary relationship and disable/enable the database.

Configuration

In ShardingSphere, cluster management can centralize rule configuration management:

- The `/rules` node saves global rule configurations, including the authority configuration of usernames and passwords in ShardingSphere-Proxy, distributed transaction type configurations, and more.
- The `/props` node stores global configuration information such as printing SQL logs and enabling cross-database queries.
- The `/metadata/${schemeName}/dataSources` node keeps data source configurations including database links, accounts, passwords, and other connection parameters.
- The `/metadata/${schemeName}/rules` node saves the rule configuration. All function rule configurations of ShardingSphere are stored under the node, such as data sharding rules, read/write splitting rules, data desensitization rules, and HA rules.
- The `/metadata/${schemeName}/schema` node stores metadata information, along with the table names, columns, and data types of logical tables.

State coordination

To share the rules and strategies of different computing nodes in the same cluster, ShardingSphere chooses a monitoring notification event mechanism to ensure such sharing. Users only need to execute SQL statements or ShardingSphere's DistSQL statements on one running instance and other running instances in the same cluster will also perceive and synchronize the operations.

To conclude, cluster management is essential for stable and correct services with HA: in terms of standalone deployment, downtime has an immeasurable impact on the system, but cluster deployment can ensure service availability.

Observability

Observability debuted in the industrial field: first, people used sensor devices to measure volumetric flow rate and substances of liquid mixtures. Then, the data was transmitted to a visual dashboard that provides operators with monitoring data, which greatly improved work efficiency.

In recent years, observability has been widespread in information technology and software systems. In particular, new IT concepts such as Cloud native, DevOps, and Intelligent ops accelerated the popularity of observability in IT. Instead of observability, in the past, people often used the concept of *monitoring*.

In particular, ops management often stresses the importance of a monitoring system. So, what is the difference between a monitoring system and observability? In simple terms, monitoring is a subset of observability. Monitoring highlights that the internal system is not known by the observer (that is, it only focuses on the *black box*), while observability emphasizes the observer's initiative and connection with a system (that is, it cares about the *white box*).

The concept of observability might be new to you, but the next section gives you its definition, followed by its usage scenarios.

Clarifying the concept of observability

What is observability? Observability refers to the characteristics of quantifiable data that a system can observe. It reflects the nature of the system itself and the real ability of the system. In IT systems, observability places its emphasis on integrating this capability into the whole system development process from design to implementation. If observability is seen as a business requirement, it should be parallel with other system development requirements such as **availability** and **scalability**.

Often, observability data is displayed via the **Application Performance Monitoring (APM)** system. The system can collect, store, and analyze observability data to perform system performance monitoring and diagnosis, and its functions include but are not limited to the performance indicator monitor, call-chain analysis, and application topology.

Applying observability to your system

In IT systems, there are three methods in which to practice observability: metrics, link tracing, and logging:

- **Metrics** leverage data aggregation to display observability and reflect the system state and trends; however, they fail to reflect system details (for example, counter, gauge, histogram, and summary).

- **Link tracing** can record data of all requests and related calls from the beginning to the end, to better show the process details.

- **Logging** logs system execution to provide detailed system operation information, but the method may cost a significant amount of resources.

In practice, system observability is shown by combining several methods. In addition, it's necessary to equip observability with a good user interface, so it is also often associated with visualization.

Mechanisms

Two mechanisms can be used to technically implement observability: one is that the internal system uses the API to provide external observable data, and the other is to collect observable data via a non-invasive probe method. The second method is quite common for developing an observability platform because its development and deployment can be independent of the system itself, which is perfect for decoupling:

Figure 4.12 – The probe method

Metrics, link tracing, and observability will certainly be of use to you at some point. With the focus now firmly placed on increasing efficiency across industries, this holds true for the desire to have an efficient database system. Observability not only allows you to gain an understanding of how your system is performing, but it will also provide you with insights to pinpoint exactly where you can improve it.

Application scenarios

With observability, you can use the data to analyze and solve online problems in IT systems. You can analyze system performance, locate slow requests, and track and audit security events. Additionally, observability can give a system action-based features such as warning and predict, to influence a user's management changes and help them make better decisions. *Figure 4.13* presents the module architecture of ShardingSphere-Proxy and its agent's plugin connection:

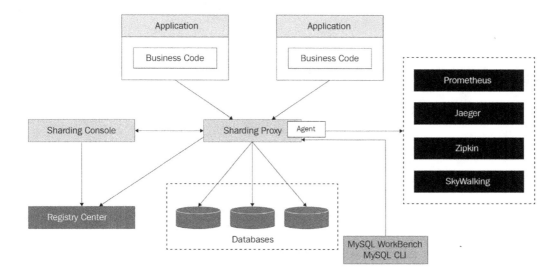

Figure 4.13 – The ShardingSphere agent module architecture

Apache ShardingSphere supports observability features, for instance, metrics, tracing, and logging. Instead of including built-in features, Apache ShardingSphere presents these features as extensible plugins so that users can develop custom plugins for Apache ShardingSphere agents and implement specific data collection, storage, and display methods. Currently, Apache ShardingSphere's default plugin supports Prometheus, SkyWalking, Zipkin, Jeager, and OpenTelemetry. Thanks to its agent module architecture, you will find that implementing observability in your version of Apache ShardingSphere is simple, and we are sure that you will find it to be a useful feature.

Summary

You have now gained a full picture of Apache ShardingSphere's features, how they are built into the system, and their use cases.

This chapter's takeaways, coupled with the knowledge that you have cumulated so far thanks to the previous three chapters, will have given you a better understanding of the architecture and the features that you can choose from. You are probably already thinking about the possible deployment options that are at your disposal and how they could help you solve your pain points.

You are now missing the last piece of the puzzle: understanding the clients that make up ShardingSphere and their deployment architectures. These are Proxy, JDBC, and hybrid deployment.

In the next chapter, we will introduce you to ShardingSphere-Proxy and ShardingSphere-JDBC deployments, along with their concurrent deployment thanks to a hybrid architecture.

5

Exploring ShardingSphere Adaptors

If you have reached this chapter after reading this book in a linear fashion, you should start to feel confident in your understanding of Apache ShardingSphere and distributed database systems.

By now, you should have acquired an overview of ShardingSphere's architecture, the features that are essential for a distributed database, and the bonus features to improve security and performance.

Next, you will be diving deeper into the workings and differences between **ShardingSphere-JDBC** and **ShardingSphere-Proxy**.

In this chapter, we will cover the following topics:

- Differences between ShardingSphere-JDBC and ShardingSphere-Proxy
- ShardingSphere-JDBC
- ShardingSphere-Proxy
- Architecture introduction

By the end of this chapter, you will be fully capable of distinguishing between ShardingSphere-JDBC and ShardingSphere-Proxy and will have clarified which one better fits your requirements. Additionally, you will learn how to deploy not only each client but also gain an understanding of the architecture and be able to deploy both simultaneously in your environment.

Technical requirements

No hands-on experience in a specific language is required but it would be beneficial to have some experience in Java since ShardingSphere is coded in Java.

To run the practical examples in this chapter, you will need the following tools:

- **JRE or JDK 8+**: This is the basic environment for all Java applications.

- **Text editor (not mandatory)**: You can use Vim or VS Code to modify the YAML configuration files.

- **A 2 cores 4 GB machine with Unix or Windows OS**: ShardingSphere can be launched on most OSs.

- **7-Zip or tar command**: You can use these tools for Linux or macOS to decompress the proxy artifact

> You can find the complete code file here:
> ```
> https://github.com/PacktPublishing/A-Definitive-
> Guide-to-Apache-ShardingSphere
> ```

Differences between ShardingSphere-JDBC and ShardingSphere-Proxy

To respond to the complicated application scenarios of distributed database ecosystems, ShardingSphere provides two independent products: ShardingSphere-JDBC and ShardingSphere-Proxy.

We could also call them **adaptors**. In terms of the overall architecture, since ShardingSphere-JDBC and ShardingSphere-Proxy use the same pluggable kernel, they can both provide standard incremental functions such as data sharding, read/write splitting, distributed transaction, and distributed governance. At the same time, with different product positioning, the two products provide two different ways of using ShardingSphere.

The following table includes a comparison between the two options:

	ShardingSphere-JDBC	ShardingSphere-Proxy
Database	Any	MySQL/PostgreSQL
Deployment method	Integrated with applications	Independent
Connected consumption	High	Low
Heterogeneous language	Java only	Any
Performance	Low loss	Relatively high loss
Decentralization	Yes	No
Static entry	Yes	No

Table 5.1 – ShardingSphere-JDBC and ShardingSphere-Proxy comparison

ShardingSphere-JDBC is positioned as a lightweight Java architecture and works with any applications written in Java. It can provide incremental services in the JDBC layer of Java. We can refer to it as an enhanced JDBC driver that can support any database achieving the JDBC standard. Since ShardingSphere-JDBC directly accesses databases via the driver, its performance loss is negligible.

ShardingSphere-Proxy is positioned as a transparent database proxy. It supports heterogeneous languages through the encapsulation of database binary protocols. Currently, ShardingSphere-Proxy provides the MySQL and PostgreSQL versions, allowing users to directly use ShardingSphere-Proxy as a database. It is suitable for any client ends that are compatible with the MySQL and PostgreSQL database protocols, making it easier for DBAs and operation and maintenance personnel to manage distributed database services.

Now, let's dive deeper into the concepts and applications of both these adaptors in the following sections.

ShardingSphere-JDBC

ShardingSphere-JDBC is the first product of ShardingSphere that succeeded, with the first open source project by the name Sharding-JDBC being released in 2017.

The long-running open source project has been continuously improved and polished thanks to the power of the community. The architectural diagram of ShardingSphere-JDBC and its application is shown as follows:

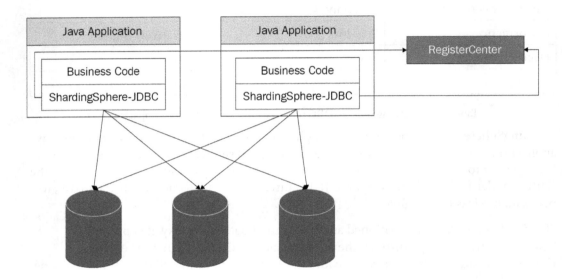

Figure 5.1 – ShardingSphere's architectural diagram

To be able to use the most mature product in the ShardingSphere ecosystem, you only need JAR files as dependencies, without an additional deployment. It is fully compatible with JDBC and ORM.

In the next section, we will look into the development mechanism of ShardingSphere-JDBC, its applicability, and the ideal users that should use it. This way, you will be able to determine whether ShardingSphere-JDBC fits your requirements. If so, you will be able to refer to the quick start guide in the *Differences between ShardingSphere-JDBC and ShardingSphere-Proxy* section.

The ShardingSphere-JDBC development mechanism

The JDBC protocol is the basis of ShardingSphere-JDBC's implementation. Here, `ShardingSphereDataSource` is created through `ShardingSphereDataSourceFactory` and implements the `DataSource` and `AutoCloseable` interfaces, ensuring that ShardingSphere-JDBC can seamlessly connect to various components that are based on the JDBC protocol.

In `ShardingSphereDataSource`, a request needs to go through the process, as highlighted by the dotted line in the following diagram of the ShardingSphere-JDBC data route – **SQL Parser | SQL Router | SQL Rewriter | SQL Executer | Result Merger**:

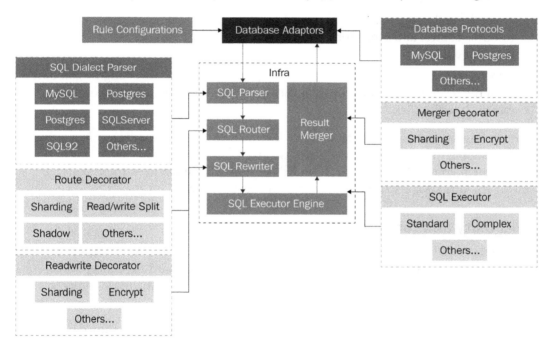

Figure 5.2 – ShardingSphere-JDBC data routing

The following list of steps details the flow of *Figure 5.2*, starting from when the SQL request is sent to when the merger engine combines the results of the execution engine:

1. First, the SQL request sent to ShardingSphere-JDBC goes through the parser engine, where it will be parsed into an **abstract syntax tree** (**AST**) by **ANother Tool for Language Recognition** (**ANTLR**).

2. Next, the router engine generates its routing path according to the context's sharding strategy.

3. At this step, the rewriter engine rewrites the logical SQL into SQL that can be executed correctly in the real database.

4. Then, the SQL request enters the execution engine. After automatically balancing resource control and execution efficiency, the engine executes the SQL request on the right database.

5. According to the results given by the execution engine, the merger engine combines all of the result sets obtained by each data node into one result set and correctly returns it to the client side.

Understanding the mechanism allows you to develop a deep understanding of *how things work*. Developing this type of knowledge is essential for building transferrable skills that will benefit you in the long term. Now, let's explore who might need these skills by reviewing the various application scenarios in the next section.

Applicability and target users

ShardingSphere-JDBC completely adheres to the JDBC standard, so you can easily get started and implement features such as read/write splitting, sharding, encryption, and distributed transactions. ShardingSphere-JDBC is ideal for making these complex functions easier.

The target users of ShardingSphere-JDBC are developers. As long as you or any developer are familiar with JDBC and JDBC-based ORM tools, you can quickly get started with ShardingSphere-JDBC.

The following list includes examples of tools that can be smoothly connected to ShardingSphere-JDBC:

*   JDBC-based ORM frameworks, such as JPA, Hibernate, Mybatis, and Spring JDBC Template

*   Third-party database connection pools including DBCP, C3P0, BoneCP, and HikariCP

*   JDBC-compliant databases, such as MySQL, PostgreSQL, Oracle, and SQL Server

Now that you are familiar with ShardingSphere's development mechanism and applicability, you might be wondering how to get started and how to deploy it.

In the following section, we provide you with a guide on the different methods that are at your disposal to deploy ShardingSphere-JDBC. Once done, you'll be ready to deploy ShardingSphere-JDBC. Note that in *Chapter 8, Apache ShardingSphere Advanced Usage – Database Plus and Plugin Platform*, we will teach you how to use it with a set of examples for the different ShardingSphere features.

Deployment and user quick start guide

It's very easy to get started with using ShardingSphere-JDBC. Perform the following steps:

1. First, we add the dependency of ShardingSphere-JDBC into the Java project:

    ```
    <dependency>
        <groupId>org.apache.shardingsphere</groupId>
        <artifactId>shardingsphere-jdbc-core</artifactId>
        <version>5.0.0</version>
    </dependency>
    ```

    > **Note**
    > You can replace the above with any version of ShardingSphere-JDBC if necessary.

2. Then, we add config files such as `resources/sharding-databases.yaml` into the path:

    ```
    mode:
      type: Standalone
      repository:
        type: File
      overwrite: true


    dataSources:
      ds_0:
        dataSourceClassName: com.zaxxer.hikari.
    HikariDataSource
        driverClassName: com.mysql.jdbc.Driver
        jdbcUrl: jdbc:mysql://127.0.0.1:13306/demo_ds?server
    Timezone=UTC&useSSL=false&useUnicode=true&character
    ```

```
Encoding=UTF-8
    username: root
    password:
  ds_1:
    dataSourceClassName: com.zaxxer.hikari.
HikariDataSource
    driverClassName: com.mysql.jdbc.Driver
    jdbcUrl: jdbc:mysql://127.0.0.1:13307/demo_ds?server
Timezone=UTC&useSSL=false&useUnicode=true&character
Encoding=UTF-8
    username: root
    password:
```

3. Next, we can use the following code to get `DataSource`:

```
DataSource dataSource =
YamlShardingSphereDataSourceFactory.
createDataSource(YamlDataSourceDemo.class.
getResource(fileName).getFile())
```

After obtaining the `DataSource` object, we can operate ShardingSphere-JDBC in the same way as any other JDBC.

In this section, we have understood the internal processes of ShardingSphere-JDBC when a database is queried. If you are interested in a JDBC driver, you will find ShardingSphere-JDBC to be very friendly and to have a very flat learning curve. It is an important adaptor of the ShardingSphere ecosystem that merits being familiar with even if you do not plan on relying on it completely, as it can be integrated with ShardingSphere-Proxy for a hybrid architecture deployment.

In the next section, will look at ShardingSphere-Proxy and follow a similar pattern to the previous section. We will start by reviewing its mechanisms and then move on to its deployment and application scenarios.

ShardingSphere-Proxy

ShardingSphere-Proxy is a database proxy that is transparent to you or any user. It provides you with all the features of the Apache ShardingSphere ecosystem, such as sharding, read/write splitting, shadow database, data encryption/decryption, and distributed governance.

Unlike ShardingSphere-JDBC, ShardingSphere-Proxy implements some of the prevailing database-access protocols and, theoretically, can support all database clients that are based on the MySQL, PostgreSQL, and openGauss protocols. When you utilize ShardingSphere-Proxy, you will feel it effectively works in the same way as a database.

The following is ShardingSphere-Proxy's deployment overview diagram:

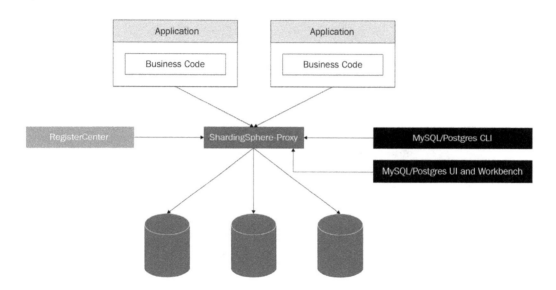

Figure 5.3 – The ShardingSphere-Proxy topography

As you can see, the proxy is positioned above multiple databases and below multiple applications. This is especially convenient if you plan on connecting multiple applications and database instances as, with the introduction of ShardingSphere-Proxy, your workflow will be greatly simplified. You won't need to deploy custom configurations for each application or database.

The ShardingSphere-Proxy development mechanism

ShardingSphere-Proxy implements MySQL, PostgreSQL, and openGauss database protocols, so it can parse the SQL and command parameters sent from the clients. Then, the ShardingSphere kernel routes and rewrites the SQL and uses the correct database driver to execute the actual SQL and parameters.

The database execution results are aggregated in ShardingSphere-Proxy and encapsulated in the protocol layer into a database network packet that is later returned to the client side, as shown in the following diagram:

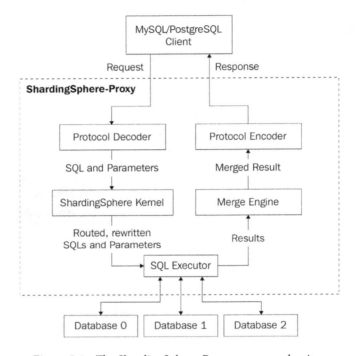

Figure 5.4 – The ShardingSphere-Proxy query mechanism

As you can see from *Figure 5.4*, because of ShardingSphere-Proxy, the SQL query is handled and directed to the correct database thanks to the SQL executor.

Now that the development mechanism is clear, let's introduce you to ShardingSphere-Proxy's applicability and target users in the next section.

Applicability and target users of ShardingSphere-Proxy

The advantages of ShardingSphere-Proxy are listed as follows:

- **Zero intrusion**: Users operate ShardingSphere-Proxy in exactly the same way as they'd operate a database.
- **Convenient for operations**: Theoretically, any client side using the MySQL or PostgreSQL databases has support access to ShardingSphere-Proxy.

- **Language-independent**: As long as a programming language is driven by a MySQL, PostgreSQL, or openGauss database, ShardingSphere-Proxy is connectable. This means that all enhanced features placed above databases in the Apache ShardingSphere ecosystem, such as data sharding and read/write splitting, are adoptable.

When it comes to possible applications and the type of user toward which ShardingSphere-Proxy is directed, there is no better way to explain it than with examples. Let's consider the following scenarios:

- **Scenario 1**: Standalone databases have held up the process of enhancing application performance. If we don't modify the application, how can we break the bottleneck?

- **Scenario 2**: An application uses ShardingSphere-JDBC to shard data and then distribute that data into 10 database instances. Now its developer or DBA wants to simply collect some data (for example, the total data volume). How can the developer/DBA do that?

- **Scenario 3**: When Python or Golang developers want to use read/write splitting, data encryption, shadow databases, or other Apache ShardingSphere features, what should they do?

Now, if we assume that there is no ShardingSphere-Proxy, users may suggest the following list of respective solutions:

- **Solution 1**: To use ShardingSphere-JDBC, the user needs to add dependencies and adjust data sources. Since the premise is that the application cannot be changed, it's necessary to find a solution at the database layer or the layer above databases.

- **Solution 2**: Users familiar with Java (or other JVM languages) can write code to operate data through ShardingSphere-JDBC, or they can use scripts to loop through all databases.

- **Solution 3**: In this case, they have to find another possible solution.

There are many other issues that are caused by ShardingSphere-JDBC's limitations as an SDK. These exact reasons led to the drivers behind the creation of ShardingSphere-Proxy. Not only can it fix problems in many scenarios where ShardingSphere-JDBC would be ineffective, but it is also friendlier for developers (of Java or other languages) and database administrators.

Deployment and user quick start guide

Users can find all of the released versions of the ShardingSphere-Proxy binary packages published on the ShardingSphere website or Docker Hub.

Additionally, users can compile ShardingSphere-Proxy by themselves by downloading the source codes of the released versions or cloning source codes of the main branch in the GitHub repository.

Taking ShardingSphere-Proxy and PostgreSQL as an example, let's see how to quickly get started.

Downloading from the official website

You can download the binary package of ShardingSphere-Proxy by clicking on the following link:

```
https://shardingsphere.apache.org/document/current/en/
downloads/
```

Here, we use version 5.0.0 for our ShardingSphere-Proxy deployment example:

1. First, download ShardingSphere-Proxy's binary package. Then, unzip it and go to the following directory:

    ```
    tar zxf apache-shardingsphere-5.0.0-shardingsphere-proxy-
    bin.tar.gz
    cd apache-shardingsphere-5.0.0-shardingsphere-proxy-bin
    ```

    Add an authority configuration in `conf/server.yaml`:

    ```
    rules:
      - !AUTHORITY
        users:
          - proxy_user@%:proxy_password
        provider:
          type: ALL_PRIVILEGES_PERMITTED
    ```

    Add data sources in `conf/config-sharding.yaml`.

    Note that no rule is required. Any SQL sent to `proxy_db` is directly routed to the configured data sources:

    ```
    schemaName: proxy_db
    dataSources:
      postgres:
        url: jdbc:postgresql://127.0.0.1:5432/postgres
        username: postgres
        password: postgres
    ```

```
      connectionTimeoutMilliseconds: 30000
      idleTimeoutMilliseconds: 60000
      maxLifetimeMilliseconds: 1800000
      maxPoolSize: 50
      minPoolSize: 1
  rules: []
```

2. Start ShardingSphere-Proxy.

   The script in the `bin` directory is used to start or end ShardingSphere-Proxy:

   ```
   bin/start.sh
   ```

3. Check whether ShardingSphere-Proxy has started successfully or not.

   You can see the proxy logs in the `logs/stdout.log` file:

   ```
   [INFO ] 2021-12-21 13:38:49.842 [main]
   o.a.s.p.i.BootstrapInitializer - Database name is
   'PostgreSQL', version is '14.0 (Debian 14.0-1.pgdg110+1)'
   [INFO ] 2021-12-21 13:38:50.026 [main] o.a.s.p.frontend.
   ShardingSphereProxy - ShardingSphere-Proxy start success
   ```

4. Connect the client side to ShardingSphere-Proxy.

   When ShardingSphere-Proxy successfully starts, you can connect a client, such as `psql`, to ShardingSphere-Proxy. The DistSQL statement, `show schema resource`, is executed as follows:

   ```
   % psql -h 127.0.0.1 -p 3307 -U proxy_user -d proxy_db
   Password for user proxy_user:
   psql (14.0 (Debian 14.0-1.pgdg110+1))
   Type "help" for help.

   proxy_db=> show schema resources;
       name   |    type    |   host    | port |    db     |
   attribute

    ----------+------------+-----------+------+-----------+---
   --------------------------------------------------------
   --------------------------
   --------------------------------------------------------
   --------------
   ```

```
postgres | PostgreSQL | 127.0.0.1 | 5432 | postgres |
{"maxLifetimeMilliseconds":1800000,"readOnly":
false,"minPoolSize":1,"idleTimeoutMilli
seconds":60000,"maxPoolSize":50,"connectionTimeout
Milliseconds":30000}
(1 row)
```

The previous steps are a perfect example of a typical ShardingSphere-Proxy deployment. As you can see, the procedure is fairly simple and straightforward.

In the following section, we will check out the ShardingSphere-Proxy source code in case you are interested in learning how to build from source and how to execute it from Docker. As these sections are optional, you can skip them and directly move on to the *Architecture introduction* section.

Building from source (optional)

If you're interested in compiling source code, please refer to the *Wiki* section of Apache ShardingSphere's GitHub repository:

`https://github.com/apache/shardingsphere/wiki`

Perform the following steps:

1. To get the source code of ShardingSphere, you can download it from `https://shardingsphere.apache.org/document/current/en/downloads/`.

    Alternatively, you can clone the source code from our GitHub repository:
    ```
    git clone --depth=1 https://github.com/apache/shardingsphere.git
    ```

2. Enter the source code directory and execute it:
    ```
    ./mvnw clean install -Prelease -T1C -DskipTests -Djacoco.skip=true -Dcheckstyle.skip=true -Drat.skip=true -Dmaven.javadoc.skip=true -B
    ```

You should be able to find the binary package of ShardingSphere-Proxy ending with `tar.gz` in the `shardingsphere-distribution/shardingsphere-proxy-distribution/target` directory.

Now you have acquired ShardingSphere-Proxy's binary package. To start it, please follow the remaining steps, as described in the *Downloading from the official website* section.

Executing ShardingSphere-Proxy with a Docker image (optional)

You can find the Docker image of ShardingSphere-Proxy in Docker Hub (`https://hub.docker.com/r/apache/shardingsphere-proxy`). Perform the following steps:

1. Get the latest ShardingSphere-Proxy Docker image, as follows:

    ```
    docker pull apache/shardingsphere-proxy:latest
    ```

2. Create the configuration directory.

    If you choose the Docker image method to use Sharding-Proxy, you need to mount the directory where the config files are located, which is at `/opt/shardingsphere-proxy/conf` in the container.

3. Create a directory to store the configuration:

    ```
    mkdir -p $HOME/shardingsphere-proxy/conf
    ```

4. Create `server.yaml` and configure the authority rules:

    ```
    rules:
      - !AUTHORITY
        users:
          - postgres@%:postgres
        provider:
          type: ALL_PRIVILEGES_PERMITTED
    ```

5. Create `config-sharding.yaml` and configure the data sources.

    Note that no rule is required. Any SQL sent to `proxy_db` is directly routed to the configured data source:

    ```
    schemaName: proxy_db
    dataSources:
      postgres:
        url: jdbc:postgresql://127.0.0.1:5432/postgres
        username: postgres
        password: postgres
        connectionTimeoutMilliseconds: 30000
        idleTimeoutMilliseconds: 60000
        maxLifetimeMilliseconds: 1800000
        maxPoolSize: 50
    ```

```
      minPoolSize: 1
rules: []
```

6. Create `logback.xml` as your log configuration.

   By default, ShardingSphere-Proxy uses `logback` as a log implementation:

   ```xml
   <configuration>
       <appender name="console" class="ch.qos.logback.core.
   ConsoleAppender">
           <encoder>
               <pattern>[%-5level] %d{yyyy-MM-dd HH:mm:ss.SSS}
   [%thread] %logger{36} - %msg%n</pattern>
           </encoder>
       </appender>
       <logger name="org.apache.shardingsphere" level="info"
   additivity="false">
           <appender-ref ref="console" />
       </logger>
       <logger name="com.zaxxer.hikari" level="error" />
       <logger name="com.atomikos" level="error" />
       <logger name="io.netty" level="error" />
       <root>
           <level value="info" />
           <appender-ref ref="console" />
       </root>
   </configuration>
   ```

7. Start ShardingSphere-Proxy:

   ```
   docker run --name shardingsphere-proxy -i -t -p3307:3307 -v
   $HOME/shardingsphere-proxy/conf:/opt/shardingsphere-proxy/
   conf apache/shardingsphere-proxy:latest
   ```

   You can now connect the PostgreSQL client to ShardingSphere-Proxy. If you see the following output, that means that you have successfully started the ShardingSphere-Proxy container:

   ```
   Starting the ShardingSphere-Proxy ...
   The classpath is /opt/shardingsphere-proxy/conf:.:/opt/
   shardingsphere-proxy/lib/*:/opt/shardingsphere-proxy/
   ext-lib/*
   ```

```
Please check the STDOUT file: /opt/shardingsphere-proxy/
logs/stdout.log

[INFO ] 2021-12-21 06:36:30.284 [main] o.a.s.p.frontend.
ShardingSphereProxy - ShardingSphere-Proxy start success
```

Architecture introduction

Real transaction scenarios, in addition to Java, always require support for additional heterogeneous languages such as Go and Python. Nevertheless, these applications using different languages still need to access and manage the same distributed database service.

Although ShardingSphere-JDBC and ShardingSphere-Proxy are two independent products, the powerful design of ShardingSphere's architecture can support the mixed deployment of the two products through one registry center, and they can be online at the same time.

Figure 5.5 provides you with a graphical representation of a deployment architecture including both ShardingSphere-JDBC and ShardingSphere-Proxy; we interchangeably refer to this type of deployment as a *mixed* or *hybrid* deployment architecture:

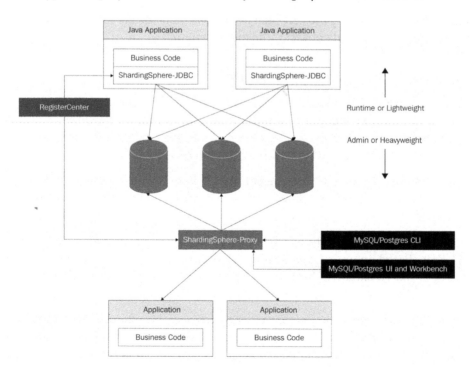

Figure 5.5 – A mixed deployment topography

As you can see in the preceding diagram, we divide the deployed architecture into two layers. When the upper layer is running, we use ShardingSphere-JDBC to deploy Java applications through a decentralized architecture, developing high-performance, lightweight OLTP applications. The administration ends up in the lower layer use of ShardingSphere-Proxy and is independent of the application deployment cluster. While providing static access and making it easier for DBAs, operations, and maintenance personnel to dynamically manage database shards, it also supports heterogeneous languages and is suitable for OLAP applications.

Between ShardingSphere-JDBC and ShardingSphere-Proxy is their mutually managed lower-layer database cluster and mutually used `RegistryCenter`. The registry center ensures that ShardingSphere-JDBC and ShardingSphere-Proxy are online at the same time. They enjoy metadata information under the same logic database. When any application modifies the metadata information in the database cluster, the registry center will push modified metadata information to all application ends in real time, ensuring the consistency of data in the whole architecture.

In mixed deployment mode, ShardingSphere-Proxy, as the administration end, cannot only modify online metadata information such as sharding rules of the whole cluster without rebooting the application but also provide features such as a one-click circuit breaker instance and disabling the secondary database. This means a more convenient and subtle operation method for database traffic management.

Applicability and target users

With the development of database ecosystems, developers have more choices in terms of database services. Services including various distributed databases, NewSQL databases, and distributed database services based on the Apache ShardingSphere distributed database middleware can all meet your needs for data services.

In comparison to other products, Apache ShardingSphere is an ecosystem consisting of multiple adaptors, and ShardingSphere-JDBC and ShardingSphere-Proxy have the same kernel architecture and open ecosystem.

In terms of features, they can both use the incremental functions provided by Apache ShardingSphere based on underlying databases. Additionally, they can use any open ecosystem-based extended function provided by ShardingSphere-JDBC and ShardingSphere-Proxy in a transparent and undifferentiated way.

Users can acquire the same function no matter which way they choose to integrate. At the same time, the needs of different application scenarios, including OLTP applications that require high performance, OLAP applications, and operations and maintenance scenarios, can all be met through ShardingSphere.

In terms of deployment, if integrated through Java applications, ShardingSphere-JDBC can easily perform transparent operations such as data sharding and read/write splitting without a separate deployment service, allowing a faster project release. At the same time, application systems suitable for different scenarios can be flexibly built by a unified, online management configuration with the mixed deployment of ShardingSphere-Proxy.

Deployment and user quick start guide

A **mixed deployment** mode requires separate deployments of the ShardingSphere-Proxy and Java applications based on ShardingSphere-JDBC.

A mixed deployment mode requires ShardingSphere-JDBC and ShardingSphere-Proxy to use the same registry center. It needs a cluster mode, which requires the separate deployment of ZooKeeper or the etcd service.

1. Add the following configuration in the ShardingSphere-Proxy `server.yaml` file and the ShardingSphere-JDBC configuration document (taking Spring Boot as an example):

   - ShardingSphere-Proxy:

     ```
     mode:
        type: Cluster
        repository:
          type: ZooKeeper
          props:
            namespace: governance_ds
            server-lists: localhost:2181
            retryIntervalMilliseconds: 500
            timeToLiveSeconds: 60
            maxRetries: 3
            operationTimeoutMilliseconds: 500
        overwrite: false
     ```

   - ShardingSphere-JDBC:

     ```
     spring.shardingsphere.mode.type=Cluster
     spring.shardingsphere.mode.repository.type=ZooKeeper
     spring.shardingsphere.mode.repository.props.namespace=
     governance_ds
     spring.shardingsphere.mode.repository.props.server-
     ```

```
lists=localhost:2181
spring.shardingsphere.mode.overwrite=false
```

> **Note**
>
> To ensure that ShardingSphere-JDBC and ShardingSphere-Proxy manage
> the same cluster, the same `namespace` and `server-lists` details are
> required in the `mode` configuration.

2. The same logic database needs to be configured in ShardingSphere-JDBC and
    ShardingSphere-Proxy:

    -   In ShardingSphere-Proxy, we can specify the logic database by
        defining `schemaname` in the YAML document under the `conf` installation
        directory. A logic database could also be dynamically created through the
        `CREATE DATABASE schemaname` statement. Here, let's take a YAML
        configuration as an example:

```
schemaName: sharding_db

dataSources:
  ds_0:
    url: jdbc:mysql://127.0.0.1:3306/demo_ds_0?
serverTimezone=UTC&useSSL=false
    username: root
    password:
    connectionTimeoutMilliseconds: 30000
    idleTimeoutMilliseconds: 60000
    maxLifetimeMilliseconds: 1800000
    maxPoolSize: 50
    minPoolSize: 1
  ds_1:
    url: jdbc:mysql://127.0.0.1:3306/demo_
ds_1?serverTimezone=UTC&useSSL=false
    username: root
    password:
    connectionTimeoutMilliseconds: 30000
    idleTimeoutMilliseconds: 60000
    maxLifetimeMilliseconds: 1800000
```

```
        maxPoolSize: 50
        minPoolSize: 1

    rules:

    ...
```

> **Note**
>
> The preceding configuration specifies that the name of a logic database managed by ShardingSphere-Proxy is `sharding_db`.

- The logic database used to configure applications in ShardingSphere-JDBC is `sharding_db`:

```
spring.shardingsphere.schema.name=sharding_db


spring.shardingsphere.mode.type=Cluster

spring.shardingsphere.mode.repository.type=ZooKeeper

spring.shardingsphere.mode.repository.props.namespace=
governance_ds

spring.shardingsphere.mode.repository.props.server-
lists=localhost:2181

spring.shardingsphere.mode.overwrite=false
```

Deployment of ShardingSphere mixed mode can be achieved by launching the ShardingSphere-JDBC and ShardingSphere-Proxy applications, respectively.

Summary

In this chapter, we introduced the two Apache ShardingSphere ecosystem adaptors: JDBC and Proxy. A detailed explanation of the differences between the two adaptors was provided, followed by a deep dive into the mechanisms of each adaptor, an example of each deployment, and the respective application scenarios.

You were also introduced to the mixed deployment of both ShardingSphere JDBC and Proxy and how they can unlock numerous possibilities. Now that you have gained an overall understanding of both ShardingSphere JDBC and Proxy, along with how to deploy them, it is time for you to dive deeper into the inner workings of each.

The following chapters, 6 and 7, will show you how to use JDBC and Proxy, respectively, to take full advantage of all the features that Apache ShardingSphere has to offer.

6
ShardingSphere-Proxy Installation and Startup

Unlike **ShardingSphere-JDBC**, **ShardingSphere-Proxy** provides a proxy service requiring independent deployment. ShardingSphere-Proxy is based on database protocols that allow users to perform operations on distributed databases in the same way as operating centralized databases – they can connect database terminals, third-party clients, or any other JDBC-supporting application to ShardingSphere-Proxy.

Now that you have reached this point of the book, you should be feeling confident in your understanding of Apache ShardingSphere and wondering how to start using it. This chapter will guide you through your first steps in installing and setting up ShardingSphere and, ultimately, getting it up and running.

In this chapter, we're going to cover the following main topics:

- Introduction to Distributed SQL
- Configuration

The first section will get you started at the beginning of the process, by showing you how to install your binary package.

Technical requirements

To run the practical examples in this chapter, you will need the following tools:

- **Text editor (not mandatory)**: You can use Vim or VS Code to modify the YAML configuration files.

- **7-Zip or tar command**: You can use these tools for Linux or macOS to decompress the proxy artifact.

- **A MySQL/PG client**: You can use the default CLI or other SQL clients such as Navicat or DataGrip to execute SQL queries.

For ShardingSphere-Proxy, you can either use the binary package or Docker to perform the installation.

> **You can find the complete code file here:**
> `https://github.com/PacktPublishing/A-Definitive-Guide-to-Apache-ShardingSphere`

Installing with the binary package

When you deploy ShardingSphere-Proxy with the binary package, please do the following:

1. Install **Java™ Runtime Environment** (**JRE**) version 8.0 or higher.

2. Download the binary package for ShardingSphere-Proxy. Navigate to `https://archive.apache.org/dist/shardingsphere/` and download the `apache-shardingsphere-${version}-shardingsphere-proxy-bin.tar.gz` file.

3. Decompress the binary package.

4. Modify the configuration files in the `conf.` directory. For the modification details, please refer to the following sections:

    - Configuration – sharding

    - Configuration – read/write splitting

    - Configuration – encryption

    - Configuration – shadow database

    - Configuration – mode

    - Configuration – scaling

    - Configuration – mixed configurations

    - Configuration – server (the `server.Yaml` configuration)

Installing with Docker

If you prefer to install with Docker, you can refer to the following steps:

1. Pull the `docker pull apache/shardingsphere-proxy` Docker image.
2. Create the `server.yaml` and `config-xxx.yaml` files under `/${your_work_dir}/conf/` to configure the services and rules.
3. To start the proxy container, please use a command similar to the following format:

    ```
    docker run -d -v /${your_work_dir}/conf:/opt/
    shardingsphere-proxy/conf -e PORT=3308 -p13308:3308
    apache/shardingsphere-proxy:latest
    ```

    > **Note – 4343**
    >
    > Users are allowed to map the container's port. For example, in the preceding command, `3308` refers to Docker's container port, while `13308` represents the host port.
    >
    > The configuration must be mounted to `/opt/shardingsphere-proxy/conf`.

At this point, the preconditions for ShardingSphere-Proxy are met.

Introduction to Distributed SQL

Leveraging the power of ShardingSphere's efficient **SQL Parser Engine**, **Distributed SQL** (**DistSQL**) is an interactive management tool developed by Apache ShardingSphere to assist users. As a part of the **distributed database management solution** (**DDBMS**), DistSQL can identify and handle resource and rule configurations and query requests.

Based on their respective functions, DistSQL can be categorized into three separate categories:

-   **Resource and Rule Definition Language** (**RDL**) is used to create, modify, and delete resources and rules.

-   **Resource and Rule Query Language** (**RQL**) is used to query resources and rules, so you won't need to check files when you want to know the configuration state.

-   **Resource and Rule Administration Language** (**RAL**) is used to control advanced features such as permission controls, transaction type switching, and circuit breakers.

However, up to now, DistSQL statements have only been supported by ShardingSphere-Proxy.

Configuration – sharding

In this section, we'd like to showcase how to configure data sharding in ShardingSphere-Proxy by means of DistSQL and YAML. With DistSQL, users can use SQL-like statements to configure and manage sharding rules.

In this section, the focus is placed on the DistSQL syntax, and some examples are given to illustrate how users can use DistSQL to perform sharding operations. Additionally, the YAML method relies on configuration files to configure and manage data sharding features. We'd like to introduce the configuration items and their definitions, as well as provide scenario examples.

DistSQL – the SQL syntax

There are four table types: the sharding table, the binding table, the broadcast table, and the single table. To perform operations such as creating table rules, modifying table rules, deleting table rules, and displaying table rules, DistSQL provides four statements respectively: CREATE, ALTER, DROP, and SHOW.

Sharding table

The DistSQL statements designed for sharding table management primarily focus on sharding rules that require a sharding strategy configuration, a sharding algorithm, and a distributed key generator. Accordingly, DistSQL statements are supported to manage sharding strategies, sharding algorithms, and distributed key generation.

This is the simplest syntax definition for creating a sharding table rule, and more optional parameters can be found in the latest documentation.

Drop sharding table rule

The syntax for drop sharding table rules is simpler than creating and modifying them. You can do so according to the following example:

```
DROP SHARDING TABLE RULE tableName [, tableName] ...
```

Query sharding table rules

Usually, you can view all sharding table rules in the current database using the SHOW SHARDING TABLE RULES syntax.

Let's look at an example. The SQL statement that is used to create a sharding table rule is shown as follows:

```
CREATE SHARDING TABLE RULE t_order (
    RESOURCES(ds_0, ds_1),
    SHARDING_COLUMN=order_id, TYPE(NAME=hash_mod,
PROPERTIES("sharding-count"=4)),
    GENERATED_KEY(COLUMN=another_id, TYPE(NAME=snowflake,
PROPERTIES("worker-id"=123)))
);
```

Binding table rules

DistSQL has also designed a related syntax for managing binding table rules. Binding table rules are based on the sharding table rules, which bind the relationships formed between different sharding table rules.

First, let's create the Sharding Binding Table Rules syntax:

```
CREATE SHARDING BINDING TABLE RULES  (tableName [, tableName]
... ) [, (tableName [, tableName]* ) ]
```

Essentially, the syntax for modifying and deleting binding table rules is the same as the creation syntax; just change CREATE to ALTER or DROP.

Next, let's query the sharding binding table rules as follows:

```
SHOW SHARDING BINDING TABLE RULES [FROM schemaName]
```

Now, let's take a look at an example. Here is the statement to create binding table rules. The primary table and the joiner table should have the same sharding rules, such as t_order table and t_order_item table. In this case, as you can see, both are using order_id as sharding key, so they are binding each other's tables. Cartesian product correlation will not appear in the multi-table correlating query, which means that the query efficiency will greatly increase. The statement is as follows:

```
CREATE SHARDING BINDING TABLE RULES (t_order, t_order_item);
```

Broadcast table

We use the DistSQL syntax for creating broadcast table rules. The broadcast table has nothing to do with the shard table rules. The table used to create the broadcast table rules can be an existing table or a table that has not been created yet.

First, let's create the `Sharding Broadcast Table Rules` syntax:

```
CREATE SHARDING BROADCAST TABLE RULES (tableName [, tableName]
...)
```

Similar to binding table rules, modifying broadcast table rules and deleting broadcast table rules only requires changing `CREATE` in the creation syntax to either `ALTER` or `DROP`.

Next, let's query `Sharding Broadcast Table Rules`:

```
SHOW SHARDING BROADCAST TABLE RULES [FROM schemaName]
```

Let's check out an example. This is the statement to create broadcast table rules:

```
CREATE SHARDING BROADCAST TABLE RULES (t_config);
```

YAML

We will now look at the YAML configurations:

- Here are the rules:

  - **Table configuration items**: The related configuration items of the table in sharding, including the ordinary sharding table, the automatic sharding table, the binding table, and the broadcast table, are mainly used to configure the relevant sharding information in order to properly run the sharding feature:

    - `Tables`: The user-defined sharding table rule configuration
    - `autoTables`: The automatic sharding table rule configuration
    - `bindingTables`: The binding table rule configuration
    - `broadcastTables`: The broadcast table rule configuration

  - **Strategy configuration items**: These are the shard-related strategy configuration items, including the default database sharding strategy, the default table sharding strategy, and the default primary key generation strategy. They can be used to specify the shard-related strategy information:

    - `defaultDatabaseStrategy`: The default database sharding strategy
    - `defaultTableStrategy`: The default table sharding strategy
    - `defaultKeyGenerateStrategy`: The default key generation strategy

- **Default sharding column**: This is the default shard key. When the sharding policy does not contain a shard key configuration, the default shard key will be used:

  - `defaultShardingColumn`: The default sharding column name

- **Sharding algorithms**: The sharding algorithm configuration information. These algorithms will calculate the corresponding routing information when sharding:

  - `ShardingAlgorithms`: The sharding algorithm configuration

- **Key generator**: This is the algorithm used to configure the primary key generation:

  - `KeyGenerators`: The key generator configuration

- **Tables**: Shard table configuration. Sharding information used to configure the table, including the table name, real data nodes, the database sharding strategy, the table sharding strategy, and the primary key generation strategy:

  - `LogicTable`: The logical table name.

  - `ActualDataNodes`: The data node name (the format is data source name + table name).

  - `DatabaseStrategy`: The database sharding strategy. Without configuration, the default database sharding strategy will be used.

  - `tableStrategy`: The table sharding strategy. Without configuration, the default table sharding strategy will be used.

  - `keyGenerateStrategy`: The key generation strategy. Without configuration, the default key generation strategy will be used.

- `autoTables`: The automatic shard table configuration. Automatic sharding information for configuring tables, including the table name, the real data source information, the sharding strategy, and the primary key generation strategy:

  - `logicTable`: The logic table name.

  - `actualDataSources`: The data source names.

  - `shardingStrategy`: The sharding strategy.

  - `keyGenerateStrategy`: The key generation strategy. Without configuration, the default key generation strategy will be used.

- `databaseStrategy`: The database sharding strategy. This is the sharding strategy for specifying data sources, including the standard sharding strategy, the complex sharding strategy, the hint sharding strategy, and the no sharding strategy:

  - `strategyType`: The strategy type, such as standard, complex, hint, and none

  - `shardingColumn`: The sharding column name

  - `shardingAlgorithmName`: The sharding algorithm name

- `tableStrategy`: The sharding strategy for the table. The user specifies the sharding strategy for the table, including the standard sharding strategy, the complex sharding strategy, the hint sharding strategy, and the no sharding strategy.

  - `strategyType`: The strategy type, such as standard, complex, hint, and none

  - `shardingColumn`: The sharding column name

  - `shardingAlgorithmName`: The sharding algorithm name

- `keyGenerateStrategy`: The primary key generation strategy configuration. With this, you can configure the primary key generation strategy under the sharding function and specify the corresponding primary key generation algorithm:

  - `column`: The column name

  - `keyGeneratorName`: The key generator name

> **Note**
>
> For the configuration items of `defaultDatabaseStrategy`, please refer to the table under `databaseStrategy`.
>
> For the configuration items of `defaultTableStrategy`, please refer to `tableStrategy`.
>
> For the configuration items of `defaultKeyGenerateStrategy`, please refer to `keyGenerateStrategy`.

- `shardingAlgorithms`: The sharding algorithm configuration. The algorithms are used to configure sharding:

  - `shardingAlgorithmName`: The sharding algorithm name

  - `type`: The sharding algorithm type

  - `props`: The sharding algorithm properties

- keyGenerators: The configuration of the primary key generation algorithm. The algorithm is used to configure the primary key generation:

  - keyGeneratorName: The key generator name

  - type: The key generator type

  - props: The key generator properties

For sharding, you will find a yaml example in the following script:

```
rules:
- !SHARDING
  tables:
    t_order:
      actualDataNodes: ds_${0..1}.t_order_${0..1}
      tableStrategy:
        standard:
          shardingColumn: order_id
          shardingAlgorithmName: t_order_inline
```

Bear in mind that you can also configure other strategies depending on your requirements.

Configuration – read/write splitting

In this section, we will introduce you to DistSQL's management syntax for read/write splitting rules:

1. First, let's create the read/write splitting rules:

```
CREATE READWRITE_SPLITTING RULE ruleName (
    WRITE_RESOURCE=resourceName,
    READ_RESOURCES(resourceName [ , resourceName] *),
    TYPE(NAME=algorithmName, PROPERTIES("key"="value" [,
"key"="value" ]* )
);
```

This is the standard syntax of the read/write splitting rule. The DistSQL syntax with read/write splitting also provides configuration methods for dynamic data sources.

To modify the syntax of the read/write splitting rule, replace CREATE with ALTER and keep the other parts unchanged.

2. Next, we will learn how to drop read/write splitting rules. The delete syntax is as follows:

```
DROP READWRITE_SPLITTING RULE ruleName, [ruleName]*
```

3. Now, let's learn how to show read/write splitting rules. The query syntax is as follows:

```
SHOW READWRITE_SPLITTING RULES [FROM schemaName]
```

The DistSQL statements used to create read/write splitting table rules are shown here:

```
CREATE READWRITE_SPLITTING RULE write_read_ds (
WRITE_RESOURCE=write_ds,
READ_RESOURCES(read_ds_0, read_ds_1),
TYPE(NAME=random)
);
```

You will find that modifying the syntax is similar to the process of creating the syntax. It is very simple to show and delete the syntax. Now, let's review the configuration options when using YAML.

YAML

Now, we will look at the configuration for the rules and data sources of ShardingSphere's read/write splitting feature:

- **Rules configuration**: You can configure various ShardingSphere rules, such as read/write splitting, data sharding, encryption, and decryption:

  - dataSources: The data source configuration

  - loadBalancers: The secondary database load balancer configuration

- **Read/write splitting data source configuration**: The primary-secondary data source information is used to configure read/write splitting to help ShardingSphere identify the primary-secondary data source:

  - writeDataSourceName: The primary database name

  - readDataSourceNames: The secondary database name

  - loadBalancerName: The secondary database load balancer configuration

In the following code, you can find an example to configure your data source for the read/write splitting feature:

```
rules:
- !READWRITE_SPLITTING
  dataSources:
    read_write_ds:
      writeDataSourceName: write_ds
      readDataSourceNames:
        - read_ds_0
        - read_ds_1
      loadBalancerName: random
  loadBalancers:
    random:
      type: RANDOM
```

This wraps up the configuration items for ShardingSphere's features. In the next section, we will take you through the encryption feature's configuration.

Configuration – encryption

Data encryption can transform data based on encryption rules and encryption algorithms to protect private user data, while data decryption reverses the data encryption process.

In this section, we will present DistSQL statements and examples concerning data encryption alongside the YAML configurations and examples.

DistSQL

Now, let's look at the DistSQL statements:

- Create Encrypt Rules:

```
CREATE ENCRYPT RULE tableName(
COLUMNS(
(NAME=columnName
,PLAIN=plainColumnName,CIPHER=cipherColumnName,
TYPE(NAME=encryptAlgorithmType
,PROPERTIES('key'='value')))
),QUERY_WITH_CIPHER_COLUMN=true);
```

To modify the syntax of the encrypt rule, replace CREATE with ALTER and keep the other parts unchanged. Here's an example:

```
CREATE ENCRYPT RULE t_encrypt (
COLUMNS(
(NAME=user_name,PLAIN=user_name_plain,
CIPHER=user_name_cipher,TYPE(NAME=AES,PROPERTIES
('aes-key-value'='123456abc'))),
(NAME=password, CIPHER =password_cipher,TYPE(NAME=MD5))
),QUERY_WITH_CIPHER_COLUMN=true);
```

- Drop Encrypt Rules: If you'd like to delete syntax, you can do so as follows:

```
DROP ENCRYPT  RULE ruleName, [ruleName]*
```

- Show Encrypt Rules: The query syntax is as follows:

```
SHOW ENCRYPT RULES [FROM schemaName]
```

Modifying the syntax is similar to creating the syntax. It is very simple to show the syntax and delete the syntax.

In actual application scenarios, you can even configure your own encryption and decryption algorithms as needed.

YAML configuration items

In Apache ShardingSphere-Proxy, the YAML configuration is also used to manage data encryption.

The following list provides the YAML configuration items for the encryption feature:

- !ENCRYPT: The encryption and decryption rule identification. This is used to identify the rule as an encryption and decryption rule:

  - tables: The encryption table configuration.

  - encryptors: The encryption algorithm configuration.

  - queryWithCipherColumn: The enable query with cipher column. It's also possible to query with a plaintext column (if there are plaintext columns).

- tables: The related configuration of encryption and decryption tables. This is the table name and corresponding column information for configuring encryption and decryption:

- `table-name`: The encryption table name
- `columns`: The encryption column configuration

- `columns`: The configuration related to the encryption and decryption columns. This is used to configure the logical names of the encryption and decryption columns and the names of the corresponding encrypted columns and auxiliary query columns, to help ShardingSphere encrypt and decrypt data correctly:

  - `column-name`: The encryption column name
  - `cipherColumn`: The cipher column name
  - `assistedQueryColumn`: The query assistant column name
  - `plainColumn`: The plaintext column name
  - `encryptorName`: The encryptor algorithm name

- `encryptors`: The configuration of encryption and decryption algorithms. Configure the relevant algorithms that can be used for encryption and decryption:

  - `encrypt-algorithm-name`: The encryption algorithm name
  - `type`: The encryption algorithm type
  - `props`: This is the encryption algorithm to configure the `assistedQueryColumn` property.

Here is an example you can refer to for your encryption:

```
rules:
- !ENCRYPT
  tables:
    t_encrypt:
      columns:
        name:
          plainColumn: name_plain
          cipherColumn: name
          encryptorName: name-encryptor
```

This is an example that you could base your potential algorithm on, but bear in mind that you can also configure other algorithms.

This marks the end of your required encryption feature configuration. Now we are going to move on to the shadow database feature configuration items.

Configuration – shadow database

Shadow databases can route SQL executions according to configured shadow rules and shadow algorithms; therefore, test data is isolated from production data. This feature can solve data corruption problems in stress tests.

In this section, we will introduce you to DistSQL statements and the examples concerning shadow databases alongside related YAML configurations and examples.

DistSQL

DistSQL also provides support for the shadow database feature. This means that you can create, modify, delete, and query shadow rules using DistSQL:

- `Create Shadow Rule`: The create syntax is as follows:

```
CREATE SHADOW RULE shadow_rule(
SOURCE=resourceName,
SHADOW=resourceName,
tableName((algorithmName,
TYPE(NAME=encryptAlgorithmType
,PROPERTIES('key'='value')))
));
```

In addition to creating shadow rules, DistSQL also provides a syntax for creating shadow algorithms and setting default shadow algorithms.

To modify the syntax of the shadow rule, replace `CREATE` with `ALTER` and keep the other parts unchanged.

- `Drop Shadow Rule`: This syntax supports removing the shadow algorithm. The delete syntax is as follows:

```
DROP SHADOW RULE ruleName, [ruleName]*
```

- `Show Shadow Rule`: This syntax also supports the query shadow algorithm. The query syntax is as follows:

```
SHOW SHADOW RULES [FROM schemaName]
```

Now, let's look at an example to create shadow database rules:

```
CREATE SHADOW RULE shadow_rule(
SOURCE=ds,SHADOW=ds_shadow,
```

```
t_order((simple_hint_algorithm, TYPE(NAME=SIMPLE_HINT,
PROPERTIES("foo"="bar")))));
```

Modifying the syntax is similar to creating the syntax. It is very simple to show and delete the syntax.

YAML

You can also use YAML configurations to configure data encryption in Apache ShardingSphere-Proxy. The following tables provide you with the configuration items for the shadow database feature:

- !SHADOW: Here are the basic configuration items and parameter descriptions of the shadow rules. Each parameter is described in detail as follows:

  - dataSources: This configures mappings from production data sources to shadow data sources.

  - tables: The shadow table name and configuration.

  - defaultShadowAlgorithmName: The default shadow algorithm name.

  - shadowAlgorithms: The shadow algorithm name and configuration.

- dataSources: Here is a list of descriptions of the data source mapping relationship configuration parameters in the shadow rule:

  - shadowDataSource: The production data source and shadow data source mapping configuration names

  - sourceDataSourceName: The production data source name

  - shadowDataSourceName: The shadow data source name

- defaultShadowAlgorithmName: Here is a description of the shadow algorithm configuration that takes effect, by default, in the shadow rule:

> **Note**
> This is optional; you can choose the configuration according to your needs.

  - defaultShadowAlgorithmName: The default algorithm name (optional)

- `tables:`
  - `tableName:` The shadow table name
  - `dataSourceNames:` The shadow-table-related data source mapping configuration list
  - `shadowAlgorithmNames:` The shadow algorithms related to the shadow tables
- `shadowAlgorithms:`
  - `shadowAlgorithmName:` The shadow algorithm name
  - `type:` The shadow algorithm type
  - `props:` The shadow algorithm configuration

Here is an example of a shadow data source mapping configuration that you can refer to:

```
dataSources:
   shadowDataSource:
#     Configure sourceDataSourceName and shadowDataSourceName
 tables:
   t_order:
#     Configure dataSourceNames and shadowAlgorithmNames
 shadowAlgorithms:
   simple_hint_algorithm:
#     Configure type and props
```

The previous example is fairly standard, and you will find that you can utilize the script yourself quite often. In the next section, we will guide you through ShardingSphere's various mode configuration items.

Configuration – mode

The mode operation is very easy for developers and testers to use, and it also contains the production environment's deployment cluster operation mode.

ShardingSphere provides three operation modes, namely **memory mode**, **local mode**, and **cluster mode**. ShardingSphere mode doesn't support modifications and changes by DistSQL. Therefore, here, we will only present the YAML configuration method.

In the following sections, you will find that we have broken down the ShardingSphere mode configuration items into `mode`, `repository`, and `props` sections:

- `mode`: With the operation mode configuration, you can configure memory, a single machine, and cluster mode. When using standalone mode or cluster mode, you can choose whether to override the remote configuration with the local configuration:

  - `type`: The operation mode: memory, standalone, and cluster
  - `repository`: The operation mode configuration
  - `overwrite`: Whether to overwrite remote configuration: `true` or `false`

- `repository`: When using the standalone mode, you can specify the persistent configuration based on the file. When using cluster mode, you can use ZooKeeper or etcd for persistent configuration information:

  - `type`: The configuration center: ZooKeeper or etcd
  - `props`: The persistent configuration

- `props`: In cluster running mode, you need to configure the namespace and registry information:

  - `namespace`: The configuration naming space
  - `server-lists`: The configuration center address

Here is an example of a mode configuration that you can refer to:

```
mode:
  type: Cluster
  repository:
    type: ZooKeeper
    props:
      namespace: governance_ds
      server-lists: localhost:2181
      retryIntervalMilliseconds: 500
      timeToLiveSeconds: 60
      maxRetries: 3
      operationTimeoutMilliseconds: 500
  overwrite: true
```

This completes our section on how to configure the various modes of Apache ShardingSphere. The next section will take you through the configuration items for the scaling feature.

Configuration – scaling

Elastic scale-out is at the core of ShardingSphere, and its relevant configurations are placed in the sharding rule YAML configuration. Although it requires a long process, it can be easily operated through DistSQL.

Scale-out has two modes, namely an automated process and a manual process. The `completionDetector` configuration should be active in automated processes, while the other configurations can be turned on according to actual needs. Then, trigger `elastic scale-out [job]` until it is done. In the manual process, each stage of elastic scale-out can be controlled.

This section will discuss two things:

- The elastic scale-out job DistSQL syntax and examples
- The elastic scale-out YAML configuration

DistSQL for job management

Users can control the whole process of scaling data migration through DistSQL, including starting and stopping scaling jobs, viewing progress, disabling reads, checking scaling, switching configurations, and more.

Currently, elastic scale-out can only be triggered by Sharding DistSQL.

The scaling syntax is different from the syntax of other features; you will find that it has more operation modes. Nevertheless, it is good for you to know the common syntax, which is as follows:

```
SHOW SCALING LIST ;
SHOW SCALING STATUS jobId ;
START SCALING jobId ;
STOP SCALING jobId ;
DROP SCALING jobId ;
RESET SCALING jobId ;
```

In addition to the previous syntax, the verification syntax is also similar:

```
CHECK SCALING jobId [BY TYPE(NAME=encryptAlgorithmType
,PROPERTIES('key'='value'))]? ;
```

You should note that scaling provides an advanced syntax, as follows:

```
SHOW SCALING CHECK ALGORITHMS ;
STOP SCALING SOURCE WRITING jobId ;
CHECKOUT SCALING jobId ;
```

The scaling process can become more complicated at times, so the provided syntax is more diverse.

Now for the examples! The following examples will prove to be handy references when it comes to scaling. Please note that because there are many scaling syntaxes, we will not show them here one by one. Only the most commonly used syntaxes are used as an example:

- Stop Scaling:

    ```
    STOP SCALING 1449610040776297;
    ```

- Start Scaling:

    ```
    START SCALING 1449610040776297;
    ```

- Stop Scaling Source Writing:

    ```
    STOP SCALING SOURCE WRITING 1449610040776297;
    ```

In the preceding example, 1449610040776297 is the operation number, which is commonly referred to as jobId. The last result returned in the example is Query OK.

YAML – configuration items

The elastic scale-out configuration is in server.yaml. The elastic scale-out job will be triggered according to the configuration when sharding needs to be adjusted.

Now, let's look at the scaling configuration items, followed by an example:

- blockQueueSize: This is the queue size of the data transmission channel.
- workerThread: This is the worker thread pool size, that is, the number of migration task threads allowed to run concurrently.

- `clusterAutoSwitchAlgorithm`: When configured, the system will detect when the scaling job has finished and switch the cluster configuration automatically.
- `dataConsistencyCheckAlgorithm`: When configured, the system will use this defined algorithm to do a data consistency check when triggered.

Here is an example of a scaling configuration that you can refer to:

```
scaling:
  blockQueueSize: 10000
  workerThread: 40
  clusterAutoSwitchAlgorithm:
    type: IDLE
    props:
      incremental-task-idle-minute-threshold: 30
  dataConsistencyCheckAlgorithm:
    type: DEFAULT
```

> **Note**
>
> Scaling for tables without primary keys, and for tables with compound primary keys is not supported.
>
> When scaling at the current storage node, a new database cluster needs to be prepared as the target end.

Configuration – multiple features, server properties

In this section, we will discuss how to realize the rule configuration of data sharding based on read/write splitting. Please note that the data source used in this process should be the one after aggregating read/write splitting.

DistSQL

The SQL syntax is consistent with each feature when used singularly. For more detailed syntax, please refer to the following sections:

- *Configuration – sharding*
- *Configuration – read/write splitting*
- *Configuration – mode*

Now for the example! Perform the following steps:

1. First, create the read/write splitting rules:

    ```
    CREATE READWRITE_SPLITTING RULE ds_0 (
        WRITE_RESOURCE=write_ds_0,
        READ_RESOURCES(write_ds_0_read_ds_0,
    write_ds_0_read_ds_1),TYPE(NAME=ROUND_ROBIN)
    ), ds_1 (
        WRITE_RESOURCE=write_ds_1,
        READ_RESOURCES(write_ds_1_read_ds_0,
    write_ds_1_read_ds_1),TYPE(NAME=ROUND_ROBIN)
    );
    ```

2. After the read/write splitting rule has been created successfully, use the read/write splitting data source as the sharded data source of the sharding table rule. Create a shard table rule for t_order:

```
CREATE SHARDING TABLE RULE t_order (
RESOURCES(ds_0,ds_1),
SHARDING_COLUMN=order_id,TYPE(NAME=hash_mod,
PROPERTIES("sharding-count"=4)),
GENERATED_KEY(COLUMN=order_id,TYPE(NAME=SNOWFLAKE,
PROPERTIES("worker-id"=123)))
);
```

Now, the mixed read/write splitting and sharding table rule creation has been completed. The corresponding rules can be viewed through the SHOW syntax.

You can also use the DDL statement to create the t_order table through the proxy, which will actually create four shard nodes on write_ds_0 and write_ds_1, that is, four tables.

YAML

The configuration items are consistent with each feature when used as a standalone feature. For detailed configuration items, please refer to the following sections:

*   *Configuration – sharding*
*   *Configuration – read/write splitting*
*   *Configuration – mode*

For example, in addition to using DistSQL to create rules, configuration files can also be used to create read/write splitting and sharding rules. The configuration files are as follows:

1. Define the logic schema and data sources:

```
schemaName: sharding_db
dataSources:
  # define data sources write_ds_0, write_ds_1,
write_ds_0_read_0, write_ds_0_read_1, write_ds_1_read_0,
write_ds_1_read_1.
```

2. Then, create the read/write splitting rule and the sharding rule in the same file:

```
readwrite_splitting
rules:
- !READWRITE_SPLITTING
# define readwrite_splitting rule here

- !SHARDING
# define sharding rule here
```

Mixed – encryption + read/write splitting + cluster

This section discusses how to create a scenario of application for both data encryption and read/write splitting.

DistSQL

The SQL syntax remains consistent with what you have learned so far. It is same as if you were configuring one single feature, instead of multiple features like we are doing here. For more detailed syntax, please refer to the following sections:

- *Configuration – encryption*
- *Configuration – read/write splitting*
- *Configuration – mode*

Let's look at an example. This section will use DistSQL to configure hybrid rules for data encryption and read/write splitting and present a complete read/write demonstration. Perform the following steps:

1. Create the read/write splitting rules:

```
CREATE READWRITE_SPLITTING RULE wr_group (
WRITE_RESOURCE=write_ds,
READ_RESOURCES(read_ds_0,read_ds_1),
TYPE(NAME=random)
);
```

2. Create the encryption rules:

```
-- Encrypt password column and store the value encrypted
in password_cipher column
CREATE ENCRYPT RULE t_encrypt (
COLUMNS(
(NAME=password,CIPHER=password_cipher,TYPE(NAME=AES,PROPE
RTIES('aes-key-value'='123456abc')))
));
```

Now, the rules for read/write splitting and encryption have been created.

3. Create an encrypted table, t_encrypt for verification.

Please note that the DDL statement only designates the original column and the extra plaintext column and that the encrypted column will be automatically processed by ShardingSphere.

The process of creating a table uses the DDL statement of Mysql, which is not shown here. However, it should be noted that the created table needs to contain the password field. The fields of the t_encrypt table are id, user_name, and password.

4. Insert data into t_encrypt:

```
-- SQL statement input
INSERT INTO t_encrypt (user_name, password) VALUES
('user_name', 'plain_password');
-- Actual SQL statement executed by ShardingSphere
INSERT INTO t_encrypt (user_name, password_cipher) VALUES
('user_name', 'OYd7QrmOWUiJKBj0oDkNIw==');
```

5. Query data in `t_encrypt`: execute the `'select * from t_encrypt'`
    statement from the proxy and the real database to verify that the data has
    been encrypted.

    Data is directly queried in a physical data source. The value of a password is
    `OYd7QrmOWUiJKBj0oDkNIw==` which shows that the password has been
    encrypted.

Data is queried in ShardingSphere. The value of the password is `plain_password`,
and the password has been automatically decrypted.

YAML configuration

Configuration items are consistent with the configuration items you have reviewed in this
chapter, under each specific feature. Concurrently configuring multiple features like we
are doing here, does not require any special or new configurations from what you learned
previously. For more detailed configuration items, please refer to the following sections:

- *Configuration – encryption*
- *Configuration – read/write splitting*
- *Configuration – mode*

In ShardingSphere-Proxy, the configuration format of the YAML file is consistent with
that of ShardingSphere-JDBC, except that `schemaName` is added to every `config-`
`xxx.yaml` file to designate the name of the logic schema. Perform the following steps:

1. Define the logic schema and data sources:

    ```
    schemaName: mixture_db


    dataSources:
      # define data sources write_ds, read_ds_0 and read_ds_1
    ```

2. Then, create the encrypt rule and the `READWRITE_SPLITTING` rule in the
    same file:

    ```
    rules:
    - !ENCRYPT
    # define encrypt rule here


    - !READWRITE_SPLITTING
    # define readwrite_splitting rule here
    ```

Compared to DistSQL, when using the YAML file for configuration, you do not need to
manually create a schema. Other operations are the same as using configured DistSQL.

Configuration – server

ShardingSphere can configure properties related to authority control, the transaction type, and the system configuration, which are all in the `server.yaml` configuration file.

Authority

ShardingSphere provides control over access to the proxy. Control over the client end's access to the proxy can be achieved through the following YAML configurations:

Name	Explanation	Default value
users	This is used to configure users who access the proxy.	None
provider.type	The authority management method; options include `ALL_PRIVILEGES_PERMITTED`/ `SCHEMA_PRIVILEGES_PERMITTED`. `ALL_PRIVILEGES_PERMITTED`: All privileges permitted by default (no authentication). `SCHEMA_PRIVILEGES_PERMITTED`: Customized authority, configured through `user-schema-mappings` properties.	None
provider.props. user-schema- mappings	This is used to specify which users can access a specific schema.	None

Table 6.1

Here's an example of the `ALL_PRIVILEGES_PERMITTED` type:

```
rules:
  - !AUTHORITY
    users:
      - root@%:root
      - sharding@:sharding
    provider:
      type: ALL_PRIVILEGES_PERMITTED
```

Here's the code for the SCHEMA_PRIVILEGES_PERMITTED type:

```
rules:
 - !AUTHORITY
   users:
     - root@:root
     - my_user@:pwd
   provider:
     type: SCHEMA_PRIVILEGES_PERMITTED
     props:
        user-schema-mappings: root@=sharding_db, root@=test_db,
my_user@127.0.0.1=sharding_db
# root users can access sharding_db when connecting on any
mainframe
# root users can access test_db when connecting on any
mainframe
# my_user users can only access sharding_db when connecting to
127.0.0.1
```

Transaction

ShardingSphere provides transaction capability. Transaction types, including LOCAL/XA/BASE, can be chosen through the YAML configuration:

Name	Explanation	Default value
defaultType	A distributed transaction used in ShardingSphere Options include LOCAL/XA/BASE	LOCAL
providerType	The realization method of the transaction specified in ShardingSphere	

Table 6.2

For example, take a look at the following:

```
rules:
 - !TRANSACTION
   defaultType: XA
   providerType: Narayana/Atomikos
```

Props configuration

ShardingSphere provides some YAML configuration items and can configure the proxy system-level properties:

- Feature switch:

Name	Explanation	Default value
sql-show	Whether to print executed SQL in log SQL	false
check-table-metadata-enabled	Whether to check the consistency of the sharding metadata structure when starting or updating the program	false
proxy-opentracing-enabled	Whether to enable OpenTracing in ShardingSphere-Proxy	false
proxy-hint-enabled	Whether to enable Hint in ShardingSphere-Proxy	false
check-duplicate-table-enabled	Whether to check duplicate tables when starting or updating the program	false
sql-federation-enabled	Whether to enable federation query	false
show-process-list-enabled	Whether to enable viewing the running SQL through SHOW PROCESSLIST only DML and DDL are included	false

Table 6.3

- Tuning parameters:

Name	Explanation	Default value
`kernel-executor-size`	This is used to set the size of the thread pool to execute tasks.	`infinite`
`max-connections-size-per-query`	The maximum connection size that can be used in every database instance per query request.	`1`
`proxy-frontend-flush-threshold`	This sets the I/O flush threshold of transmitting data rows in ShardingSphere-Proxy.	`128`
`proxy-backend-query-fetch-size`	The rows of data fetched each time the proxy backend interacts with the database (with a cursor).	`-1`
`proxy-frontend-executor-size`	The proxy frontend Netty thread pool size. The default value of `0` refers to using the Netty default value.	`0`
`proxy-backend-executor-suitable`	Options include `OLAP` and `OLTP`.	`OLAP`
`proxy-frontend-max-connections`	The maximum user end connections to the proxy; the default value of `0` refers to no limitation.	`0`

Table 6.4

Here's an example:

```
props:
    max-connections-size-per-query: 1
    kernel-executor-size: 16
    proxy-frontend-flush-threshold: 128
    proxy-opentracing-enabled: false
    proxy-hint-enabled: false
    sql-show: true
    check-table-metadata-enabled: false
    show-process-list-enabled: false
    proxy-backend-query-fetch-size: -1
    check-duplicate-table-enabled: false
```

```
sql-comment-parse-enabled: false
proxy-frontend-executor-size: 0
proxy-backend-executor-suitable: OLAP
proxy-frontend-max-connections: 0
sql-federation-enabled: false
```

That concludes our transaction feature's configuration items, giving you a complete overview of how to configure this feature in your system.

Summary

Now, you are ready to take the necessary steps to get your version of Apache ShardingSphere up and running. Thanks to this chapter, you now understand how to download and install ShardingSphere, which is an important milestone toward achieving your distributed database goals.

You can consider yourself adept at using Apache ShardingSphere. In the next chapter, you will be able to take your skillset to an advanced level by mastering ShardingSphere-JDBC, too.

Mastering ShardingSphere-JDBC will set you up to be able to fully leverage the Database Plus concept and the plugin-oriented platform that sets this ecosystem apart.

7

ShardingSphere-JDBC Installation and Start-Up

If you have read *Chapter 5, Exploring ShardingSphere Adaptors,* or you are somewhat familiar with JDBCs, you may know that they are fairly straightforward and simple to include in your system.

Using **ShardingSphere-JDBC** won't require you to prepare any additional deployment or services, as you will only need to input the dependencies and configurations into the project. This chapter helps you do just that.

Thanks to this chapter, you will be able to set up and get started with using ShardingSphere-JDBC. Moreover, we have prepared a few bonuses that will provide you with the necessary items to be able to prepare your own custom sharding strategies, configuration, and more.

In this chapter, we are going to cover the following topics:

- Setup and configuration
- Configurations

By the end of the chapter, you'll be ready to start leveraging ShardingSphere-JDBC to take your system to the next level. In the first section, we will start by introducing you to the requirements and the configuration method.

Technical requirements

For ShardingSphere-JDBC, you can use Maven to retrieve the dependencies.

Setup and configuration

To get started, we will go through a two-step procedure. The first part will help you ensure that you have correctly prepared all the requirements, while the second part will give you an overview of the configuration method.

Introducing the preliminary requirements

At the very beginning, if you are looking to use ShardingSphere-JDBC, first add a ShardingSphere-JDBC dependency. Take the following Maven dependency as an example:

```
<dependencies>
  <dependency>
    <groupId>org.apache.shardingsphere</groupId>
    <artifactId>shardingsphere-jdbc-core</artifactId>
    <version>5.0.0</version>
  </dependency>
</dependencies>
```

Then, create a ShardingSphere-JDBC configuration file. If you choose a Java configuration, skip this step. Take the `config-sharding.yaml` YAML configuration as an example. First, we need to define `mode` and `dataSources`:

```
mode:
  type: Standalone
  repository:
    type: File
  overwrite: true

dataSources:
  ds_0:
```

```
      dataSourceClassName: com.zaxxer.hikari.HikariDataSource
      driverClassName: com.mysql.jdbc.Driver
      jdbcUrl: jdbc:mysql://localhost:3306/demo_
ds_0?serverTimezone=UTC&useSSL=false&useUnicode=true&character
Encoding=UTF-8
      username: root
      password:
   ds_1:
      dataSourceClassName: com.zaxxer.hikari.HikariDataSource
      driverClassName: com.mysql.jdbc.Driver
      jdbcUrl: jdbc:mysql://localhost:3306/demo_ds_1?
serverTimezone=UTC&useSSL=false&useUnicode=
true&characterEncoding=UTF-8
      username: root
      password:
```

Then, we define `rules` and `props`, as follows:

```
rules:
- !SHARDING
  tables:
    t_order:
      actualDataNodes: ds_${0..1}.t_order
      keyGenerateStrategy:
        column: order_id
        keyGeneratorName: snowflake

  defaultDatabaseStrategy:
    standard:
      shardingColumn: user_id
      shardingAlgorithmName: database_inline

  shardingAlgorithms:
    database_inline:
      type: INLINE
      props:
        algorithm-expression: ds_${user_id % 2}
```

```
keyGenerators:
    snowflake:
        type: SNOWFLAKE
        props:
            worker-id: 123

props:
  sql-show: false
```

Now, you've successfully configured the preliminary setup.

Introducing the configuration method

As we mentioned in the introduction, you will find that configuring ShardingSphere-JDBC is very easy. ShardingSphere-JDBC supports the following four configuration methods:

- **Java API**: The Java API lays the foundation for all of ShardingSphere-JDBC's configurations because all other methods are actually internally converted into the Java API method through code. This method is perfect in scenarios that require dynamic configuration.

- **YAML**: YAML files can greatly simplify configurations. To use ShardingSphere-JDBC, the most common method is writing a configuration into its configuration file.

- **Spring Boot Starter**: ShardingSphere's Spring Boot Starter makes it easy for developers to get started with ShardingSphere-JDBC in the Spring Boot project.

- **Spring namespace**: By leveraging namespaces and dependencies provided by ShardingSphere, developers such as yourself can quickly use ShardingSphere-JDBC in Spring projects.

Let's now move on to learn how to configure specific features to be used with ShardingSphere-JDBC. The next section will get you started with data sharding and will be followed by other sections, introducing notable features such as read/write splitting, data encryption, a shadow database, and cluster mode.

Sharding configurations

In this section, the sharding configurations are provided to help you quickly understand sharding capabilities. ShardingSphere-JDBC provides four configuration methods, and you can choose one of them to quickly access a system.

Java configuration items

This part will show some configuration items related to sharding.

In this section's following tables, we will see some specific configuration items for sharding rule configuration.

The main class is `ShardingRuleConfiguration`:

Name	Data type	Note	Default value
tables (+)	Collection<ShardingTable RuleConfiguration>	Sharding table rule collection	-
autoTables (+)	Collection<ShardingAuto TableRuleConfiguration>	Auto-sharding table rule collection	-

Table 7.1

The following table gives you the `binding` table configuration:

Name	Data type	Note	Default value
bindingTableGroups (*)	Collection<String>	Binding table rule collection	None

Table 7.2

This is about `broadcast` table configuration:

Name	Data type	Note	Default value
broadcastTables (*)	Collection<String>	Broadcast table rule collection	None

Table 7.3

This is about strategy configuration:

Name	Data type	Note	Default value
defaultDatabase ShardingStrategy (?)	ShardingStrategy Configuration	Default database sharding strategy	No sharding
defaultTable ShardingStrategy (?)	ShardingStrategy Configuration	Default table sharding strategy	No sharding
defaultDatabase ShardingStrategy (?)	ShardingStrategy Configuration	Default database sharding strategy	No sharding

Table 7.4

This is about sharding columns and algorithms:

Name	Data type	Note	Default value
defaultSharding Column (?)	String	Default sharding column name	None
sharding Algorithms (+)	Map<String, ShardingSphere AlgorithmConfiguration>	Sharding algorithm name and configuration	None

Table 7.5

This is about key generators:

Name	Data type	Note	Default value
keyGenerators (?)	Map<String, ShardingSphere AlgorithmConfiguration>	Key generation algorithm name and configuration	None

Table 7.6

The following tables show you the specific configuration fields for the *table-level sharding* rules.

The main class is `ShardingTableRuleConfiguration`, and the following table shows the sharding table configuration:

Name	Data type	Note	Default value
`logicTable`	`String`	Logic table name	-

Table 7.7

This is the actual data node configuration:

Name	Data type	Note	Default value
`actualData Nodes (?)`	`String`	Data source name and table name (separated by decimal points), multi-tables (comma-separated), and support row expression	Use known data sources and logical table names to generate data nodes. This is applied to broadcast tables, or a situation where sharding databases without table sharding have exactly the same table structure.

Table 7.8

In the following table, you will find the sharding strategy configuration:

Name	Data type	Note	Default value
`databaseSharding Strategy (?)`	`ShardingStrategy Configuration`	Database sharding strategy	Use the default database sharding strategy.
`tableSharding Strategy (?)`	`ShardingStrategy Configuration`	Table sharding strategy	Use the default table sharding strategy.
`keyGenerate Strategy (?)`	`KeyGenerator Configuration`	Key generator	Use the default key generator.

Table 7.9

The following table shows the specific configuration items of the auto-sharding tables.

The main class is `ShardingAutoTableRuleConfiguration`:

Name	Data type	Note	Default value
logicTable	String	Sharding logical table name	-
actualData Sources (?)	String	Data source names (comma-separated)	Use all configurations' data sources.

Table 7.10

This is the strategy configuration:

Name	Data type	Note	Default value
logicTable	String	Sharding logical table name	-
actualData Sources (?)	String	Data source names (comma-separated)	Use all configurations' data sources.

Table 7.11

The following table presents you with the standard sharding strategy configuration items.

The main class is `StandardShardingStrategyConfiguration`:

Name	Data type	Note
shardingColumn	String	Sharding column name
shardingAlgorithmName	String	Sharding algorithm name

Table 7.12

This table is about standard complex sharding strategy configuration items.

The main class is `ComplexShardingStrategyConfiguration`:

Name	Data type	Note
shardingColumns	String	Sharding column names (comma-separated)
shardingAlgorithmName	String	Sharding algorithm name

Table 7.13

This table shows `hint` sharding strategy configuration items.

The main class is `HintShardingStrategyConfiguration`:

Name	Data type	Note
shardingAlgorithmName	String	Sharding algorithm name

Table 7.14

If you configure a `none` sharding strategy, then there is nothing else that you would need to configure.

The main class is `NoneShardingStrategyConfiguration`.

The following table shows the distributed key generation strategy configuration items.

The main class is `KeyGenerateStrategyConfiguration`:

Name	Data type	Note
column	String	Distributed key generation column name
keyGeneratorName	String	Distributed key generation algorithm name

Table 7.15

Now that we have looked at the various Java configuration items, let's look at a few examples. This example shows you how to create a sharding rule using the Java API:

```
public DataSource getDataSource() throws SQLException {
    return ShardingSphereDataSourceFactory.
createDataSource(createModeConfiguration(), createDataSourceMap(),
Collections.singleton(createShardingRuleConfiguration()), new
Properties());
}
```

The following example shows how to create a sharding rule configuration:

```
private ShardingRuleConfiguration
createShardingRuleConfiguration() {
    ShardingRuleConfiguration result = new
ShardingRuleConfiguration();
    result.getTables().add(getOrderTableRuleConfiguration())
    result.getKeyGenerators().put("snowflake", new
ShardingSphereAlgorithmConfiguration("SNOWFLAKE",
```

```
getProperties()));
   return result;
}
```

This one shows how to create a mode configuration:

```
private static ModeConfiguration createModeConfiguration() {
   return new ModeConfiguration("Standalone", new
StandalonePersistRepositoryConfiguration("File", new
Properties()), true);
}
```

Lastly, the following example shows how to create a table rule configuration:

```
private static ShardingTableRuleConfiguration
getOrderTableRuleConfiguration() {
   ShardingTableRuleConfiguration result = new
ShardingTableRuleConfiguration("t_order");
   result.setKeyGenerateStrategy(new
KeyGenerateStrategyConfiguration("order_id", "snowflake"));
   return result;
}
```

YAML configuration items

This part will show sharding configuration items in YAML that you can configure.

The following list shows `tables` that you can configure:

- `tables`: This is the data sharding rule configuration logical table name. The following items can be configured as follows:

  - `logic-table-name`: Logical table name.

  - `actualDataNodes`: Data source names and table names (please refer to inline syntax rules).

  - `databaseStrategy`: Database sharding strategy. Without other configurations, use the default database sharding strategy.

  - `tableStrategy`: Table sharding strategy.

  - `keyGenerateStrategy`: Distributed key generation strategy.

- `autoTables`: Auto-sharding table rule configuration. Here, you can configure the following items:

  - `logic-table-name`: Logical table name.

  - `actualDataNodes`: Data source name.

  - `shardingStrategy`: Database sharding strategy. Without other configurations, use the default database sharding strategy.

- `bindingTables`: Binding table rule collection. For this, we have the following:

  - `logic-table-name`: Logical table name collection

- `broadcastTables`: Broadcast table rule collection. We have the following:

  - `table-name`: Logical table name collection

This list introduces `Strategy` that you can configure:

- `defaultDatabaseStrategy`: Default database sharding strategy:

  - `strategyType`: Strategy type – `standard`, `complex`, `hint`, or `none`

  - `shardingColumn`: Sharding column name

  - `shardingAlgorithmName`: Sharding algorithm name

  Remember that `defaultDatabaseStrategy` is the same as `databaseStrategy`.

- `defaultTableStrategy`: Default table sharding strategy

- `defaultKeyGenerateStrategy`: Default distributed key generation strategy:

  - `column`: Key generation column. Without new configurations, the key generator is not enabled by default.

  - `keyGeneratorName`: Distributed key generation algorithm name.

Remember that `defaultKeyGenerateStrategy` is the same as `keyGenerateStrategy`.

Then, we have a default sharding column, sharding algorithms, and key generators that you can configure:

- `defaultShardingColumn`: Default sharding column name

- `shardingAlgorithms`: Sharding algorithm configuration

- `sharding-algorithm-name`: Sharding algorithm name

- `type`: Sharding algorithm type

- `props`: Sharding algorithm property configuration

- `keyGenerators`: Distributed key generation algorithm configuration:

  - `key-generate-algorithm-name`: Distributed key generation algorithm name

  - `type`: Distributed key generation algorithm type

  - `props`: Distributed key generation algorithm property configuration

Now, let's look at a demo of YAML about a sharding table rule:

```
tables:
   t_order:
      actualDataNodes: ds_${0..1}.t_order
      keyGenerateStrategy:
         column: order_id
         keyGeneratorName: snowflake
```

Let's take a look at an example about strategy:

```
defaultDatabaseStrategy:
   standard:
      shardingColumn: user_id
      shardingAlgorithmName: database-inline
```

And lastly, here is an example with the algorithms:

```
shardingAlgorithms:
   database-inline:
      type: INLINE
      props:
         algorithm-expression: ds_${user_id % 2}
```

Spring Boot configuration items

This part shows the items that you can configure in `springboot`:

- The first table and configuration we introduce are about actual data nodes:

  `spring.shardingsphere.rules.sharding.tables.<table-name>.actual-data-nodes`

  This has a data source name and a table name (separated by decimal points). Multiple table names are separated by commas; inline expressions are supported. By default, the system uses known data sources and logical table names to generate data nodes; it is applied to broadcast tables or situations where sharding databases without table sharding have exactly the same table structure.

- This is the sharding column name:

  - Sharding column name: `spring.shardingsphere.rules.sharding.tables.<table-name>.database-strategy.standard.sharding-column`

  - Sharding column name (comma-separated): `spring.shardingsphere.rules.sharding.tables.<table-name>.database-strategy.complex.sharding-columns`

- This is the algorithm name:

  - Sharding algorithm name: `spring.shardingsphere.rules.sharding.tables.<table-name>.database-strategy.standard.sharding-algorithm-name:`

  - Sharding algorithm name: `spring.shardingsphere.rules.sharding.tables.<table-name>.database-strategy.complex.sharding-algorithm-name:`

  - Sharding algorithm name: `spring.shardingsphere.rules.sharding.tables.<table-name>.database-strategy.hint.sharding-algorithm-name`

- This is about strategy:

  - Table sharding strategy – the same as the database sharding strategy: `spring.shardingsphere.rules.sharding.tables.<table-name>.table-strategy.xxx`

- This is about key generation:

  - Distributed key generation column name: `spring.shardingsphere.rules.sharding.tables.<table-name>.key-generate-strategy.column`
  - Distributed key generation algorithm name: `spring.shardingsphere.rules.sharding.tables.<table-name>.key-generate-strategy.key-generator-name`

- This is about bind tables:

  - Binding table rules collection: `spring.shardingsphere.rules.sharding.binding-tables[x]:`

- This is about broadcast:

  - Broadcast table rules collection: `spring.shardingsphere.rules.sharding.broadcast-tables[x]`

- This is about algorithms:

  - Sharding algorithm type: `spring.shardingsphere.rules.sharding.sharding-algorithms.<sharding-algorithm-name>.type`
  - Sharding algorithm property configuration: `spring.shardingsphere.rules.sharding.sharding-algorithms.<sharding-algorithm-name>.props.xxx`

Let's look at the following code snippet example, which provides a demo of sharding in Spring Boot:

```
spring.shardingsphere.rules.sharding.tables.t_order.actual-data-nodes=ds-$->{0..1}.t_order_$->{0..1}
spring.shardingsphere.rules.sharding.tables.t_order.table-strategy.standard.sharding-column=order_id
spring.shardingsphere.rules.sharding.tables.t_order.table-strategy.standard.sharding-algorithm-name=t-order-inline
```

SpringNameSpace configuration items

This part shows the configuration items that you can configure in `SpringNameSpace`.

Each table shows the configuration items for different rules, as follows:

- Sharding rules:

Name	Type	Note
table-rules (?)	Label	Sharding table rule configuration
auto-table-rules (?)	Label	Auto-sharding table rule configuration
binding-table-rules (?)	Label	Binding table rule configuration
broadcast-table-rules (?)	Label	Broadcast table rule configuration

Table 7.16

- Table rules:

Name	Type	Note
actual-data-nodes	Property	Data source name and table name (separated by a decimal point), multiple table name (comma-separated), and support for inline expressions
database-strategy-ref	Property	Standard database sharding strategy name
table-strategy-ref	Property	Standard table sharding strategy name

Table 7.17

- Binding table rules:

Name	Type	Note
binding-table-rule (+)	Label	Binding table rule configuration
logic-tables	Property	Binding table name (comma-separated)

Table 7.18

- Broadcast table rules:

Name	Type	Note
broadcast-table-rule (+)	Label	Broadcast table rule configuration
table	Property	Broadcast table name

Table 7.19

- Standard strategy:

Name	Type	Note
id	Property	Standard sharding strategy name
sharding-column	Property	Sharding column name
algorithm-ref	Property	Sharding algorithm name

Table 7.20

- Complex strategy:

Name	Type	Note
id	Property	Complex sharding strategy name
sharding-columns	Property	Sharding column name (comma-separated)
algorithm-ref	Property	Sharding algorithm name

Table 7.21

- Hint strategy:

Name	Type	Note
id	Property	Hint sharding strategy name
algorithm-ref	Property	Sharding algorithm name

Table 7.22

- None strategy:

Name	Type	Note
id	Property	Sharding strategy name

Table 7.23

- Key generate strategy:

Name	Type	Note
id	Property	Distributed key generation strategy name
column	Property	Distributed key generation column name
algorithm-ref	Property	Distributed key generation algorithm name

Table 7.24

- Sharding algorithm:

Name	Type	Note
id	Property	Sharding algorithm name
type	Property	Sharding algorithm type
props (?)	Label	Sharding algorithm property configuration

Table 7.25

- Key generate algorithm:

Name	Type	Note
id	Property	Distributed key generation algorithm name
type	Property	Distributed key generation algorithm name
props (?)	Label	Distributed key generation property configuration

Table 7.26

Now, let's look at an example of a demo of `SpringNameSpace` about sharding rules:

```
<sharding:rule id="shardingRule">
    <sharding:table-rules>
        <sharding:table-rule logic-table="t_order" actual-
data-nodes="demo_ds_${0..1}.t_order_${0..1}" database-strategy-
ref="databaseStrategy" table-strategy-ref="orderTableStrategy"
key-generate-strategy-ref="orderKeyGenerator" />
    </sharding:table-rules>
</sharding:rule>
```

Understanding read/write splitting configuration

In this section, the read/write splitting configuration is explained to help you quickly understand its related functions. ShardingSphere-JDBC provides you with four configuration methods, and you can freely choose one of them.

Java configuration items

The tables in this section will show you the configuration items related to read/write splitting that you can configure.

- The main class is `ReadwriteSplittingRuleConfiguration`:

Name	Data type	Note
dataSources (+)	Collection<ReadwriteSplitting DataSourceRuleConfiguration>	Read/write data source configuration
loadBalancers (*)	Map<String, ShardingSphere AlgorithmConfiguration>	Secondary database and load balancer algorithm configuration

Table 7.27

In the following table, you will see the primary-secondary data source configuration for the read/write splitting feature, with their respective names, data types, and a description for each:

- The main class is `ReadwriteSplittingDataSourceRuleConfiguration`:

Name	Data type	Note	Default
name	String	Read/write splitting data source name	-
autoAwareData SourceName (?)	String	Auto-discovered data source name (used with database discovery)	-
writeData SourceName	String	Write-only data source name	-
readData SourceNames (+)	Collection <String>	Read-only data source name collection	-
loadBalancer Name (?)	String	Read-only load balancer algorithm name	Load balancer algorithm polling

Table 7.28

The following code block provides you with a demonstration that you can refer to for configuring read/write splitting in Java:

```
ReadwriteSplittingDataSourceRuleConfiguration dataSourceConfig
= newReadwriteSplittingDataSourceRuleConfiguration(
            "demo_read_query_ds", "", "demo_write_ds",
Arrays.asList("demo_read_ds_0", "demo_read_ds_1"), "demo_
weight_lb");

    Properties algoritProperties = new Properties();
    algoritProperties.put("demo_read_ds_0", "2");
    algoritProperties.put("demo_read_ds_1", "1");
    ShardingSphereAlgorithmConfiguration algorithmConfiguration
= newShardingSphereAlgorithmConfiguration("WEIGHT",
algoritProperties);
```

YAML configuration items

In the list in this section, we will show you the read/write splitting items that you can configure in YAML:

- This shows you how to configure the read/write split rule:

    - `dataSources`: Data source of read/write splitting

    - `loadBalancers`: Load balance algorithm configuration

- This is about data source configuration:

    - `data-source-name`: Logic data source name of read/write splitting

    - `autoAwareDataSourceName`: Auto-aware data source name (use with database discovery)

    - `writeDataSourceName`: Write data source name

    - `readDataSourceNames`: Read data source names and multiple data source names separated with commas

    - `loadBalancerName`: Load balance algorithm name

- This is about load balancing:

    - `load-balancer-name`: Load balance algorithm name

    - `type`: Load balance algorithm type

    - `props`: Load balance algorithm properties

This code block provides you with a demo of read/write splitting configuration in YAML:

```
dataSources:
  pr_ds:
    writeDataSourceName: write_ds
    readDataSourceNames: [read_ds_0, read_ds_1]
    loadBalancerName: weight_lb
loadBalancers:
  weight_lb:
    type: WEIGHT
    props:
      read_ds_0: 2
      read_ds_1: 1
```

Spring Boot configuration items

In the tables in this section, we will show you the items that you can configure in Spring Boot:

- This table shows the data source:

  - Auto-discovered data source name (used with database discovery): `spring.shardingsphere.rules.readwrite-splitting.data-sources.<readwrite-splitting-data-source-name>.auto-aware-data-source-name`

  - Write-only data source name: `spring.shardingsphere.rules.readwrite-splitting.data-sources.<readwrite-splitting-data-source-name>.write-data-source-name`

  - Read-only data source names (comma-separated): `spring.shardingsphere.rules.readwrite-splitting.data-sources.<readwrite-splitting-data-source-name>.read-data-source-names`

- This is about `loadbalancer`:

  - Load balancer algorithm name: `spring.shardingsphere.rules.readwrite-splitting.data-sources.<readwrite-splitting-data-source-name>.load-balancer-name`

- Load balancer algorithm type: `spring.shardingsphere.rules.readwrite-splitting.load-balancers.<load-balance-algorithm-name>.type`

- Load balancer algorithm property configuration: `spring.shardingsphere.rules.readwrite-splitting.load-balancers.<load-balance-algorithm-name>.props.xxx`

This code block provides you with a demo of read/write splitting configuration in Spring Boot:

```
Spring.shardingsphere.rules.readwrite-splitting.load-balancers.
round_robin.type=ROUND_ROBIN
spring.shardingsphere.rules.readwrite-splitting.data-sources.
pr_ds.write-data-source-name=write-ds
spring.shardingsphere.rules.readwrite-splitting.data-sources.
pr_ds.read-data-source-names=read-ds-0,read-ds-1
spring.shardingsphere.rules.readwrite-splitting.data-sources.
pr_ds.load-balancer-name=round_robin
<readwrite-splitting:load-balance-algorithm id="randomStrategy"
type="RANDOM" />
<readwrite-splitting:rule id="readWriteSplittingRule">
    <readwrite-splitting:data-source-rule id="demo_ds"
write-data-source-name="demo_write_ds"read-data-source-
names="demo_read_ds_0, demo_read_ds_1" load-balance-algorithm-
ref="randomStrategy" />
</readwrite-splitting:rule>
```

SpringNameSpace configuration items

In this section, we will show you the items that you can configure in SpringNameSpace:

- This is about the read/write splitting rule – `<readwrite-splitting:rule />`:

Name	Type	Note
id	Property	Spring Bean Id
data-source-rule (+)	Label	Read/writing splitting data source rule configuration

Table 7.29

- This is about the data source rule – `<readwrite-splitting:data-source-rule />`:

Name	Type	Note
`id`	Property	Read/write splitting data source rule name
`auto-aware-data-source-name`	Property	Auto-discovered data source name (used with database discovery)
`write-data-source-name`	Property	Write-only data source name
`read-data-source-names`	Property	Read-only data source names (comma-separated)
`load-balance-algorithm-ref`	Property	Load balancer algorithm name

Table 7.30

- This is about the load balance algorithm – `<readwrite-splitting:load-balance-algorithm />`:

Name	Type	Note
`id`	Property	Load balance algorithm name
`type`	Property	Load balance algorithm type
`props (?)`	Label	Load balancer algorithm property configuration

Table 7.31

A SpringNameSpace example

This part provides a demo of the read/write splitting rule in `SpringNameSpace`:

```
<readwrite-splitting:load-balance-algorithm id="randomStrategy"
type="RANDOM" />
<readwrite-splitting:rule id="readWriteSplittingRule">
    <readwrite-splitting:data-source-rule id="demo_ds"
write-data-source-name="demo_write_ds"read-data-source-
names="demo_read_ds_0, demo_read_ds_1" load-balance-algorithm-
ref="randomStrategy" />
</readwrite-splitting:rule>
```

Understanding data encryption configuration

In this section, the data encryption configuration is provided to help you quickly understand the related functions. ShardingSphere-JDBC provides four configuration methods, and you can choose the appropriate one to access the system.

Java configuration items

This section gives you the configuration items related to encrypt rules that you can configure:

- The main class is `EncryptRuleConfiguration`:

Name	Data type	Note	Default
tables (+)	Collection <EncryptTable RuleConfiguration>	Encryption table rule configuration.	-
encryptors (+)	Map<String, ShardingSphere Algorithm Configuration>	Encryption algorithm name and configuration.	-
queryWith CipherColumn (?)	boolean	Configure to query with cipher columns or notes. If there is a plaintext column, it's okay to query with that.	TRUE

Table 7.32

- This is about `encrypt` column rule configuration. The main class is `EncryptTableRuleConfiguration`:

Name	Data type	Note
name	String	Table name.
columns (+)	Collection<Encrypt ColumnRuleConfiguration>	Encryption column rule configuration collection.
queryWithCipher Column (?)	boolean	Decide whether the table enables a query with a cipher column.

Table 7.33

- This is about `cipher` column rule configuration. The main class is `EncryptColumnRuleConfiguration`:

Name	Data source	Note
`logicColumn`	`String`	`logical` column name

Table 7.34

- The `config` column – this table presents you with items to configure the encryption columns' names:

Name	Data source	Note
`cipherColumn`	`String`	`cipher` column name
`assistedQueryColumn` `(?)`	`String`	`query` `assistant` column name
`plainColumn` `(?)`	`String`	`plaintext` column name

Table 7.35

- The `encrypt` name – the following table provides you with the configuration for the encryption algorithm:

Name	Data source	Note
`encryptorName`	`String`	Encryption algorithm name

Table 7.36

- This is about encryption algorithm configuration. The following table introduces you to the encryption algorithm configuration items, having `ShardingSphereAlgorithmConfiguration` as the main class:

Name	Data type	Note
`name`	`String`	Encryption algorithm name
`type`	`String`	Encryption algorithm type
`properties`	`Properties`	Encryption algorithm property configuration

Table 7.37

This code snippet provides a demo of `encrypt` rules in Java:

```
EncryptColumnRuleConfiguration columnConfigAes = new
EncryptColumnRuleConfiguration("user_name", "user_name", "",
```

```
"user_name_plain", "name_encryptor");
    EncryptTableRuleConfiguration encryptTableRuleConfig
= new EncryptTableRuleConfiguration("t_user", Arrays.
asList(columnConfigAes, columnConfigTest), null);
    encryptAlgorithmConfigs.put("name_encryptor", new
ShardingSphereAlgorithmConfiguration("AES", props));
```

YAML configuration items

This part shows the configuration items that you can configure:

- The following list shows the encrypt rule items:

  - `tables`: Encryption table configuration.

  - `encryptors`: Encryption algorithm configuration.

  - `queryWithCipherColumn`: Enable a query with a cipher column or not. If there is a plaintext column, it's okay to query with that.

- The following list shows the table configuration items:

  - `table-name`: Encryption table name

  - `columns`: Encryption column configuration

- Next, you will find the columns' configuration items:

  - `column-name`: Encrypted column name

  - `cipherColumn`: Cipher column name

  - `assistedQueryColumn`: Query assistant column name

  - `plainColumn`: Plaintext column name

- The following list introduces you to the encryption name configuration items, as well as the encryptors' configuration items:

  - `encryptorName`: Encryption algorithm name

  - `encrypt-algorithm-name`: Encryption algorithm name

  - `type`: Encryption algorithm type

  - `props`: Encryption algorithm property configuration – query assistant column names

This code block provides you with a demo of the encrypt rule in YAML:

```
t_user:
   columns:
      user_name:
         plainColumn: user_name_plain
         cipherColumn: user_name
         encryptorName: name-encryptor
```

This is about encryptors:

```
encryptors:
name-encryptor:
   type: AES
   props:
      aes-key-value: 123456abc
```

Spring Boot configuration items

This part shows you the configuration items in Spring Boot:

- Enable a query with a `cipher` column or not: `spring.shardingsphere.rules.encrypt.tables.<table-name>.query-with-cipher-column`
- `cipher` column name: `spring.shardingsphere.rules.encrypt.tables.<table-name>.columns.<column-name>.cipher-column`
- query column name: `spring.shardingsphere.rules.encrypt.tables.<table-name>.columns.<column-name>.assisted-query-column`
- `plaintext` column name: `spring.shardingsphere.rules.encrypt.tables.<table-name>.columns.<column-name>.plain-column`

The following list introduces you to the algorithm configuration items in Spring Boot:

- Encryption algorithm name: `spring.shardingsphere.rules.encrypt.tables.<table-name>.columns.<column-name>.encryptor-name`
- Encryption algorithm type: `spring.shardingsphere.rules.encrypt.encryptors.<encrypt-algorithm-name>.type`
- Encryption algorithm property configuration: `spring.shardingsphere.rules.encrypt.encryptors.<encrypt-algorithm-name>.props.xxx`

And finally, this gives you the cipher query configuration items with Spring Boot:

- Enable a query with a `cipher` column or not. If there is a `plaintext` column, it's okay to query with that: `spring.shardingsphere.rules.encrypt.queryWithCipherColumn`

This part shows the configuration about `encrypt` in Spring Boot:

```
spring.shardingsphere.rules.encrypt.tables.t_user.columns.user_
name.cipher-column=user_name
spring.shardingsphere.rules.encrypt.tables.t_user.columns.user_
name.encryptor-name=name-encryptor
```

SpringNameSpace configuration items

This part will show the configuration items related to `encrypt` in `SpringNameSpace`.

This table shows encrypt rule configuration items. Let's first start with the rules by looking at `<encrypt:rule />`:

Name	Type	Note	Default
id	Property	Spring Bean Id	
queryWith CipherColumn (?)	Property	Enable a query with a cipher column or not. If there is a plaintext column, it's okay to query with that.	TRUE
table (+)	Label	Encryption table configuration.	

Table 7.38

Next, we can move to the table configuration items with `<encrypt:table />`:

Name	Type	Note
name	Property	Encryption table name.
column (+)	Label	Cipher column configuration.
query-with-cipher-column(?)	Property	Enable a query with a cipher column or not. If there is a plaintext column, it's okay to query with that.

Table 7.39

The columns also have their dedicated configuration items with `<encrypt:column />`:

Name	Type	Note
`logic-column`	Property	Cipher column's logical name
`cipher-column`	Property	Cipher column name
`assisted-query-column (?)`	Property	Query assistant column name
`plain-column (?)`	Property	Plaintext column name
`encrypt-algorithm-ref`	Property	Encryption algorithm name

Table 7.40

Finally, let's look at the encryption algorithm's configuration items with
`<encrypt:encrypt-algorithm />`:

Name	Type	Note
`id`	Property	Encryption algorithm name
`type`	Property	Encryption algorithm type
`props (?)`	Label	Encryption algorithm property configuration

Table 7.41

This part provides a demo of the encrypt rule in `SpringNameSpace`:

```
<encrypt:rule id="encryptRule">
    <encrypt:table name="t_user">
        <encrypt:column logic-column="user_name" cipher-
column="user_name" plain-column="user_name_plain" encrypt-
algorithm-ref="name_encryptor" />
    </encrypt:table>
</encrypt:rule>
```

Configuring a shadow database

In this section, we will discuss how to use the rule configuration of a shadow database. When using `Hint` algorithms, you also need to turn on the `sqlCommentParseEnabled` `SQL_PARSER`-related configuration item to `true`.

Java configuration items

Here is the configuration item entry – org.apache.shardingsphere.shadow.
api.config.ShadowRuleConfiguration:

Name	Data type	Explanation
dataSources	Map<String, ShadowData SourceConfiguration>	Production data source and shadow data source reflection configuration
tables	Map<String, ShadowTable Configuration>	Shadow database name and configuration
defaultShadow AlgorithmName	String	Default shadow algorithm name
shadowAlgorithms	Map<String, ShardingSphere AlgorithmConfiguration>	Shadow algorithm name and configuration

Table 7.42

As an example, let's create a data source with ShadowRule using Java code:

```
public DataSource getDataSource() throws SQLException {
    Map<String, DataSource> dataSourceMap = createDataSourceMap();
    Collection<RuleConfiguration> ruleConfigurations =
createRuleConfiguration();
    return ShardingSphereDataSourceFactory.
createDataSource(dataSourceMap, ruleConfigurations, properties);
}
```

YAML configuration items

If you are using YAML to configure your shadow database feature, you can refer to
!SHADOW and the following list's configuration items:

- dataSources: Shadow database logic data source reflection configuration list
- tables: Shadow table configuration list
- defaultShadowAlgorithmName: Default shadow algorithm name (optional)
- shadowAlgorithms: Shadow algorithm configuration list

Here is a YAML configuration example for shadow data sources:

```
rules:
- !SHADOW
  dataSources:
    shadowDataSource:
      sourceDataSourceName: ds
      shadowDataSourceName: ds_shadow
```

Here are the configuration examples for shadow tables:

```
tables:
    t_user:
        dataSourceNames:
            - shadowDataSource
        shadowAlgorithmNames:
            - user_id_insert_value_match-algorithm
```

Here are the configuration examples for shadow algorithms:

```
shadowAlgorithms:
    user_id_insert_value_match-algorithm:
        type: VALUE_MATCH
        props:
            operation: insert
            column: user_id
            value: 1
```

A Spring Boot example

This section introduces you to configuring ShardingSphere's Shadow DB feature while using Spring Boot. If you use Spring Boot, you can refer to the following example code.

Note that to be able to incorporate this code example, you will need to replace boilerplate values such as ds, ds_shadow, and user_id:

```
spring.shardingsphere.rules.shadow.data-sources.shadow-data-
source.source-data-source-name=ds
spring.shardingsphere.rules.shadow.data-sources.shadow-data-
source.shadow-data-source-name=ds_shadow
spring.shardingsphere.rules.shadow.tables.t_user.data-source-
```

```
names=shadow-data-source
spring.shardingsphere.rules.shadow.tables.t_user.shadow-
algorithm-names=user_id_insert_value_match-algorithm
spring.shardingsphere.rules.shadow.shadow-algorithms.user-id-
insert-match-algorithm.type=VALUE_MATCH
spring.shardingsphere.rules.shadow.shadow-algorithms.user-id-
insert-match-algorithm.props.operation=insert
spring.shardingsphere.rules.shadow.shadow-algorithms.user-id-
insert-match-algorithm.props.column=user_id
spring.shardingsphere.rules.shadow.shadow-algorithms.user-id-
insert-match-algorithm.props.value=1
```

SpringNameSpace configuration items

The following table provides you with the shadow database configuration items for
SpringNameSpace:

Name	Type	Explanation
<shadow:data-source />	Properties	Shadow data source name
<shadow:shadow-table />	Properties	Shadow tables
<shadow:algorithm />	Properties	Shadow algorithms

Table 7.43

Let's now look at a SpringNameSpace example, including the SpringNameSpace
configuration items that we introduced in the previous table:

```
<shadow:shadow-algorithm id="user-id-insert-match-algorithm"
type="VALUE_MATCH">
    <props>
        <prop key="operation">insert</prop>
        <prop key="column">user_id</prop>
        <prop key="value">1</prop>
    </props>
</shadow:shadow-algorithm>
<shadow:rule id="shadowRule">
    <shadow:data-source id="shadow-data-source" source-data-
source-name="ds" shadow-data-source-name="ds_shadow"/>
    <shadow:shadow-table name="t_user" data-sources="shadow-
```

```
data-source">
        <shadow:algorithm shadow-algorithm-ref="user-id-
insert-match-algorithm" />
    </shadow:shadow-table>
</shadow:rule>
<shardingsphere:data-source id="shadowDataSource" data-source-
names="ds,ds_shadow" rule-refs="shadowRule">
</shardingsphere:data-source>
```

Configuring ShardingSphere's modes

This section provides the basic operations' mode configurations. In addition to cluster deployment in a production scenario, a corresponding operation mode such as standalone mode for development and automation testing scenarios is also provided for engineers. The three operation modes provided by Apache ShardingSphere are memory mode, standalone mode, and cluster mode. Memory mode is not covered here, as it is the standard auto mode that ShardingSphere uses to run.

Java configuration items

The tables in this section will show you the configuration items for modes that you can configure.

The main class is ModeConfiguration:

Name	Data type	Note
type	String	Type of mode configuration. Values can be Memory, Standalone, and Cluster.
repository(?)	Persist Repository Configuration	Persist repository configuration. The Memory type does not need to persist and can be null. The Standalone type uses StandalonePersistRepository Configuration. The Cluster type uses ClusterPersistRepository Configuration.
overwrite(?)	boolean	Overwrite persistent configuration with local configuration.

Table 7.44

For Standalone mode, the main class is
`StandalonePersistRepositoryConfiguration`:

Name	Data type	Note	Default
type	String	Type of persist repository. Values can be `File`.	-
props(?)	Properties	Properties of the persist repository. The properties key can be `path`.	`.shardingsphere`

Table 7.45

When it comes to Cluster mode, the main class is
`ClusterPersistRepositoryConfiguration`:

Name	Data type	Note	Default
type	String	Type of persist repository. Values can be `ZooKeeper` and `Etcd`.	-
namespace	String	Namespace of the registry center.	-
serverLists	String	Server lists of the registry center.	-
props(?)	Properties	Properties of the persist repository.	-

Table 7.46

A Java example

This part provides you with a demo of modes in Java.

The first example we provide here is code that you can refer to for configuring Standalone mode:

```
Standalone Mode
private ModeConfiguration getModeConfiguration(final
ShardingType shardingType) {
    StandalonePersistRepositoryConfiguration standaloneConfig
= newStandalonePersistRepositoryConfiguration("File", new
Properties());
    return new ModeConfiguration("Standalone", standaloneConfig,
true);
}
```

To configure Cluster mode, you can refer to the following code:

```
private ModeConfiguration getModeConfiguration(final
ShardingType shardingType) {
    ClusterPersistRepositoryConfiguration
clusterPersistRepositoryConfiguration =
newClusterPersistRepositoryConfiguration("ZooKeeper",
"governance", "127.0.0.1", newProperties());
    return new ModeConfiguration("Cluster",
clusterPersistRepositoryConfiguration, true);
}
```

YAML configuration items

This section provides the configuration items to be used in case you prefer to utilize YAML:

- **Memory Mode**:

  - mode (?): #: Default value is Memory

  - type: #: Type of mode configuration whose values can be Memory, Standalone, and Cluster

  - repository (?): Persist repository configuration

  - Memory type does not need persist overwrite: #: Whether to overwrite persistent configuration with local configuration

- **Standalone Mode**:

  - mode: type: #: Type of mode configuration

  - repository type: #: Type of persist repository

  - props: #: Properties of persist repository

  - path: #: Persist configuration path

  - overwrite: #: Whether to overwrite persistent configuration with local configuration

- **Cluster Mode**:

  - `mode: type:` #: Type of mode configuration
  - `repository: type:` #: Type of persist repository whose values can be `ZooKeeper` and `etcd`
  - `props:` #: Properties of persist repository
  - `namespace:` #: Namespace of registry center
  - `server-lists:` #: Server lists of registry center
  - `overwrite:` #: Whether to overwrite persistent configuration with local configuration

Now that you are aware of the configuration items, we provide you with a demo of mode configuration with YAML.

As with the previous example, we will start off with an example for `Standalone` mode:

```
Standalone Mode
mode:
  type: Standalone
  repository:
    type: File
    props: Properties of persist repository
      path:
  overwrite: true
```

Next, you can refer to the following example to configure `Cluster` mode with YAML:

```
// Cluster Mode
mode:
  type: Cluster
  repository:
    type: ZooKeeper
    props:
      namespace: governance
      server-lists: localhost:2181
  overwrite: true
```

Spring Boot configuration items

This subsection introduces you to the configuration items to be used if you prefer Spring Boot:

- The following list introduces the main mode configuration items:

  - `spring.shardingsphere.mode.type`: Type of mode configuration. Values can be Memory, Standalone, and Cluster.

  - `spring.shardingsphere.mode.repository`: Persist repository configuration. Memory type does not need to persist and can be `null`. The Standalone type uses `StandalonePersistRepositoryConfiguration`. The Cluster type uses `ClusterPersistRepositoryConfiguration`.

  - `spring.shardingsphere.mode.overwrite`: Whether to overwrite persistent configuration with local configuration.

- The following list introduces the configuration items for `Standalone` mode:

  - `spring.shardingsphere.mode.type`: Type of mode configuration.

  - `spring.shardingsphere.mode.repository.type`: Type of persist repository. Values can be `File`.

  - `spring.shardingsphere.mode.repository.props.path`: Properties of the persist repository. The properties key can be `path`.

  - `spring.shardingsphere.mode.overwrite`: Whether to overwrite persistent configuration with local configuration.

- The following list introduces the configuration items relevant to `Cluster` mode:

  - `spring.shardingsphere.mode.type`: Type of mode configuration.

  - `spring.shardingsphere.mode.repository.type`: Type of persist repository. Values can be `ZooKeeper` and `etcd`.

  - `spring.shardingsphere.mode.repository.props.namespace`: Namespace of the registry center.

  - `spring.shardingsphere.mode.repository.props.server-lists`: Server lists of the registry center.

  - `spring.shardingsphere.mode.repository.props.<key>=`: Properties of the persist repository.

  - `spring.shardingsphere.mode.overwrite`: Whether to overwrite the persistent configuration with local configuration.

This part provides you with a demo of configuring modes with Spring Boot.

The first example illustrates how to configure `Standalone` mode:

```
// Standalone Mode
spring.shardingsphere.mode.type=Standalone
spring.shardingsphere.mode.repository.type=File
spring.shardingsphere.mode.repository.props.path=
spring.shardingsphere.mode.overwrite=true
```

The second example presents you with code to configure `Cluster` mode:

```
// Cluster Mode
spring.shardingsphere.mode.type=Cluster
spring.shardingsphere.mode.repository.type=Zookeeper
spring.shardingsphere.mode.repository.props.namespace=governance
spring.shardingsphere.mode.repository.props.server-
lists=localhost:2181
spring.shardingsphere.mode.overwrite=true
```

A SpringNameSpace example

This part provides a demo of configuring mode in `SpringNameSpace`. The first example introduces Standalone mode, while the second example introduces how to configure Cluster mode:

```
<!-- Standalone Mode -->
<shardingsphere:mode type="Standalone" repository-ref="standalone
Repository"overwrite="true"/>
<standalone:repository id="standaloneRepository" type="File">
    <props>
            <prop key="path"></prop>
    </props>
</standalone:repository>
<!-- Cluster Mode -->
<shardingsphere:mode type="Cluster" repository-
ref="clusterRepository" overwrite="true"/>
<cluster:repository id="clusterRepository" type="ZooKeeper"
namespace="governance" server-lists="localhost:2181">
    <props>
```

```
        <prop key="max-retries">3</prop>
        <prop key="operation-timeout-milliseconds">3000</prop>
    </props>
  </cluster:repository>
```

Props configuration for JDBC

This section introduces the props configuration for JDBC access, which includes optimization parameters for ShardingSphere internal functions and some dynamic switch configurations. With the optimization parameters in the props configuration, you can flexibly tune to achieve optimal performance on the JDBC access side, while with the dynamic switch configuration, you can quickly locate issues and improve problem-solving efficiency.

Java configuration items

Let's follow our usual flow and start with the Java configuration items first. You will find the items classified by common, optimized, or checked props:

- **Common props**: In the following table, the SQL props are presented with their types, descriptions, and default values. These will allow you to set the SQL properties:

Name	Data type	Description	Default value
sql-show	boolean	Whether to print SQL in the log	false
sql-simple	boolean	Whether to print simple style SQL in the log	false
sql-federation-enabled	boolean	Whether to turn on federation queries	false

Table 7.47

- **Optimized props**: The following table shows you the optimization props. These will allow you to set the task-processing thread pool executor size and the maximum number of connections allowed for a single query request:

Name	Data type	Description	Default value
`kernel-executor-size`	`int`	For setting the task-processing `ThreadPoolExecutor` size	0
`max-connections-size-per-query`	`int`	The maximum number of connections allowed for a single query request	1

Table 7.48

- **Checked props**: This table details the configuration items for checked props to check shared metadata or duplicate tables:

Name	Data type	Description	Default value
`check-table-metadata-enabled`	`boolean`	Whether the structural consistency of the shared metadata is checked	`false`
`check-duplicate-table-enabled`	`boolean`	Whether the duplicate table is checked when the program starts and updates	`false`

Table 7.49

Now that we have looked at the configuration items, let's use them in an example, as follows:

```Java
public DataSource getDataSource() throws SQLException {
    Properties props = new Properties();
    props.put("sql-show", false);
    // Add more props
    return ShardingSphereDataSourceFactory.
createDataSource(createDataSourceMap(), Collections.emptyList()
props);
}
```

YAML configuration items

Please refer to the preceding section in this chapter.

The following code is an example of YAML:

```YAML
props:
  sql-show: false
  # Add more props
```

Spring Boot configuration items

Please refer to the *common props table* in the preceding *Java configuration items* section. The configuration items are the same.

The following snippet is an example of Spring Boot:

```SQL
spring.shardingsphere.props.sql-show=false
  # Add more props
```

SpringNameSpace configuration items

Please refer to all the *Java configuration items* sections in this chapter.

The following snippet is an example of SpringNameSpace:

```XML
<shardingsphere:data-source id="shardingDataSource" data-source-names="ds_0, ds_1" rule-refs="shardingRule">
    <props>
        <prop key="sql-show">false</prop>
        <!-- Add more props -->
    </props>
</shardingsphere:data-source>
```

Configuration – miscellaneous

This section discusses how to achieve rule configuration of data sharding based on read/write splitting. Please note that the sharding data source should be the one aggregated after read/write splitting.

Sharding, read/write splitting, and cluster configuration items

Configuration items are consistent with those used in each feature. For detailed configuration items, please refer to the following sections in this chapter:

- *Sharding configuration*
- *Understanding read/write splitting configuration*
- *Configuring ShardingSphere's modes*

The following example introduces the combination of sharding, read/write splitting, and Cluster mode configuration items. These steps will show you how to combine configurations for multiple features:

1. Create a sharding configuration using the Java API:

```
public final class ShardingConfigurationCreator {
    public static ShardingRuleConfiguration create() {
        // Create sharding rule configuration , please
refer to sharding.
    }
}
```

2. Create a read/write splitting configuration using the Java API:

```
public final class ReadwriteSplittingConfigurationCreator {
    public static ReadwriteSplittingRuleConfiguration
create() {
        // Create read write splitting rule configuration
, please refer to read/write splitting.
    }
}
```

3. Create a mode configuration using the Java API:

```
public final class ModeConfigurationCreator {
    public static ModeConfiguration create() {
        // Create mode configuration , please refer to
cluster mode.
    }
}
```

4. Add the created sharding configuration, read/write splitting configuration, and mode configuration to the ShardingSphere data source:

```
public final class
ShardingReadwriteSplittingClusterConfigurationCreator {
    public DataSource create() throws SQLException {
        return ShardingSphereDataSourceFactory.
createDataSource(
                    ModeConfigurationCreator.create(),
createDataSourceMap(),
                    Arrays.
asList(ShardingConfigurationCreator.create(),
ReadwriteSplittingConfigurationCreator.create()),
createProperties());
    }
}
```

Now that you have understood the basic concept for a multi-feature configuration, let's look at the different examples of configuration file structures in the following subsections.

A YAML example

The yaml file structure for configuring sharding and read/write splitting is as follows:

```
# Configure cluster mode, please refer to cluster mode.
mode:
 type: Cluster
# Configure the required data source
dataSources:
- !SHARDING
# Configure sharding rules, please refer to sharding.
- !READWRITE_SPLITTING
# Configure read/write splitting rules, please refer to read/
write splitting.
```

A Spring Boot example

The `properties` file structure for configuring sharding and read/write splitting is as follows:

```
# Configure cluster mode, please refer to cluster mode.
spring.shardingsphere.mode.type=Cluster
# Configure the required data source
spring.shardingsphere.datasource....
# Configure sharding rules, please refer to sharding.
spring.shardingsphere.rules.sharding.tables....
# Configure read/write splitting rules, please refer to read/
write splitting.
spring.shardingsphere.rules.readwrite-splitting.data-sources....
```

A SpringNameSpace example

The `xml` file structure for configuring sharding and read/write splitting is as follows:

```
# Configure cluster mode, please refer to cluster mode.
<shardingsphere:mode />
  # Configure the required data source
  <bean id="demo_write_ds_0" />
  # Configure read/write splitting rules, please refer to read/
write splitting.
  <readwrite-splitting:rule id="readWriteSplittingRule" />
  # Configure sharding rules, please refer to sharding.
  <sharding:table-rules>
```

Configuring sharding, encryption, and cluster mode

This section discusses how to use data sharding and data encryption properties together. Please note that you only need to use the logic table name for configuration when writing encryption rules.

Configuration items are consistent with those used in each feature. For detailed configuration items, please refer to the following sections in this chapter:

- *Sharding configurations*
- *Understanding data encryption configuration*
- *Configuring ShardingSphere's modes*

A Java example

Let's look at how to implement this multi-configuration with Java first. In the following steps, you can refer to this example to understand the logic to combine the configurations of multiple features:

1. Create a sharding configuration using the Java API:

```
public final class ShardingConfigurationCreator {
    public static ShardingRuleConfiguration create() {
        // Create sharding rule configuration, please refer
    to sharding.
    }
}
```

2. Create a data encryption configuration using the Java API:

```
public final class EncryptRuleConfigurationCreator {
    public static EncryptRuleConfiguration create() {
        // Create data encryption rule configuration ,
    please refer to encryption.
    }
}
```

3. Create a mode configuration using the Java API:

```
public final class ModeConfigurationCreator {
    public static ModeConfiguration create() {
        // Create mode configuration, please refer to
    cluster mode.
    }
}
```

4. Add the created sharding configuration, encryption configuration, and mode configuration to the ShardingSphere data source:

```
public final class
ShardingEncryptionClusterConfigurationCreator {
    public DataSource create() throws SQLException {
        return ShardingSphereDataSourceFactory.
    createDataSource(
                    ModeConfigurationCreator.create(),
    createDataSourceMap(),
```

```
                        Arrays.
asList(ShardingConfigurationCreator.create(),
EncryptRuleConfigurationCreator.create()),
createProperties());
    }
```

Now that you have understood the basic principle for configuring sharding and data encryption together, we can move on to see their implementation with YAML, Spring Boot, and SpringNameSpace.

A YAML example

The yaml file structure for configuring sharding and encryption is as follows:

```
# Configure cluster mode, please refer to cluster mode.
mode:
 type: Cluster
# Configure the required data source
dataSources:
- !SHARDING
# Configure sharding rules, please refer to sharding.
- !ENCRYPT
# Configure encrypt rules, please refer to data encryption.
```

A Spring Boot example

The properties file structure for configuring sharding and encryption is as follows:

```
# Configure cluster mode, please refer to cluster mode.
spring.shardingsphere.mode.type=Cluster
# Configure the required data source,
spring.shardingsphere.datasource....
# Configure sharding rules, please refer to sharding.
spring.shardingsphere.rules.sharding.tables....
# Configure encryption rules, please refer to data encryption.
spring.shardingsphere.rules.encypt.tables....
```

SpringNameSpace example

The `xml` file structure for configuring sharding and encryption is as follows:

```
# Configure cluster mode, please refer to cluster mode.
<shardingsphere:mode />
 # Configure the required data source
 <bean id="ds_0" />
 # Configure encryption rules, please refer to data encryption.
 <encrypt:rule id="encryptRule">
# Configure sharding rules, please refer to sharding.
 <sharding:table-rules>
```

Depending on your preferred method, you can easily combine the configuration of multiple ShardingSphere features simultaneously, as we saw from the previous examples.

Summary

Thanks to this chapter, you have now learned how to configure ShardingSphere-JDBC in multiple ways.

Depending on whether you prefer to work with Java, YAML, SpringNameSpace, or Spring Boot, you are now fully equipped to configure all of the features to match your preferences and requirements.

Understanding this chapter means that you have now mastered how to configure both ShardingSphere-Proxy and JDBC, giving you the power to either choose the client that you prefer or to deploy both for a hybrid architecture.

You may be wondering if that is too advanced, and that is exactly where the next chapter will take you – ShardingSphere's advanced usage and Database Plus.

Section 3: Apache ShardingSphere Real-World Examples, Performance, and Scenario Tests

On completion of this part, you will walk away feeling empowered thanks to the thorough understanding you'll have developed of using ShardingSphere. This part will provide you with the chance to apply theory to practice through the scenarios and real-world examples provided and to find out how to get the best performance out of your system thanks to baseline and performance testing.

This section comprises the following chapters:

- *Chapter 8, Apache ShardingSphere Advanced Usage – Database Plus and Plugin Platform*
- *Chapter 9, Baseline and Performance Test System Introduction*
- *Chapter 10, Testing Frequently Encountered Application Scenarios*
- *Chapter 11, Exploring the Best Use Cases for ShardingSphere*
- *Chapter 12, Applying Theory to Practical Real-World Examples*
- *Appendix and the Evolution of the Apache ShardingSphere Open Source Community*

8

Apache ShardingSphere Advanced Usage – Database Plus and Plugin Platform

This chapter will take your understanding of, and capability in leveraging, ShardingSphere to the next level. After covering the basic features and configurations, you will learn how to customize ShardingSphere according to your requirement, to get the most out of your system.

We will cover the following topics in this chapter:

- Plugin platform introduction and SPI
- User-defined functions and strategies – SQL parser, sharding, read/write splitting, and distributed transactions

- User-defined functions and strategies – data encryption, SQL authority, user authentication, SQL authority, shadow DB, and distributed governance

- ShardingSphere-Proxy – tuning properties and user scenarios

By the end of this chapter, you will be able to create, configure, and run a custom version of ShardingSphere for distributed transactions, read/write splitting, and other features.

Technical requirements

No hands-on experience in a specific language is required but it would be beneficial to have some experience in Java since ShardingSphere is coded in Java.

To run the practical examples in this chapter, you will need the following tools:

- **JRE or JDK 8+**: This is the basic environment for all Java applications.

- **Text editor (not mandatory)**: You can use Vim or VS Code to modify the YAML configuration files.

- **An IDE**: You can use tools such as Eclipse or IntelliJ IDEA for coding.

- **A MySQL/PG client**: You can use the default CLI or other SQL clients such as Navicat or DataGrip to execute SQL.

- **A 2 cores 4 GB machine with a Unix OS or Windows OS**: ShardingSphere can be launched on most OSs.

You can find the complete code file here:

```
https://github.com/PacktPublishing/A-Definitive-
Guide-to-Apache-ShardingSphere
```

Important Note

The *Introducing Database Plus* section is an extract from an excellent article written by Zhang Liang (one of the authors of this book) that explains what Database Plus is all about. You can read the complete article here: `https://faun.pub/whats-the-database-plus-concepand-what-challenges-can-it-solve-715920ba65aa` (License: CC0).

Introducing Database Plus

Database Plus is our community's design concept for distributed database systems, designed to build an ecosystem on top of fragmented heterogeneous databases. With this concept, our goal is to provide globally scalable and enhanced computing capabilities, while at the same time maximizing the original database computing capabilities.

With the introduction of this concept, the interaction between applications and databases is oriented toward the Database Plus standard, which means that the impact of database fragmentation is greatly minimized when it comes to upper-layer services. We set out to pursue this concept with the three keywords that define it: Connect, Enhance, and Pluggable.

ShardingSphere's pursuit of Database Plus

To understand how we have leveraged these three keywords to formulate our development concept, we thought that an introduction to each one of them would be helpful to you. Let us dive deeper into each one.

Connect – building upper-level standards for databases

Rather than providing an entirely new standard, Database Plus provides an intermediate layer that can be adapted to a variety of SQL dialects and database access protocols. This implies that an open interface to connect to various databases is provided by ShardingSphere.

Thanks to the implementation of the database access protocol, Database Plus provides the same experience as a database and can support any development language, and database access client.

Moreover, Database Plus supports maximum conversion between SQL dialects. An AST (abstract syntax tree) that parses SQL can be used to regenerate SQL according to the rules of other database dialects. The SQL dialect conversion makes it possible for heterogeneous databases to access each other. This way, users can use any SQL dialect to access heterogeneous underlying databases.

ShardingSphere's database gateway is the best interpretation of Connect. It is the prerequisite for Database Plus to provide a solution for database fragmentation. This is done by building a common open docking layer positioned in the upper layer of the database, to pool all the access traffic of the fragmented databases.

Enhance – database computing enhancement engine

Following decades of development, databases now boast their own query optimizer, transaction engine, storage engine, and other time-tested storage and computing capabilities and design models. With the advent of the distributed and cloud-native era, the original computing and storage capabilities of the database will be scattered and woven into a distributed and cloud-native level of new capabilities.

Database Plus adopts a design philosophy that emphasizes traditional database practices, while at the same time adapting to the new generation of distributed databases. Whether centralized or distributed, Database Plus can repurpose and enhance the storage and native computing capabilities of a database.

The capabilities enhancement mainly refers to three aspects: distributed, data control, and traffic control.

Data fragmentation, elastic scaling, high availability, read/write splitting, distributed transactions, and heterogeneous database federated queries based on the vertical split are all capabilities that Database Plus can provide at the global level for distributed heterogeneous databases. It doesn't focus on the database itself, but on the top of the fragmented database, focusing on the global collaboration between multiple databases.

In addition to distributed enhancement, data control and traffic control enhancements are both in the silo structure. Incremental capabilities for data control include data encryption, data desensitization, data watermarking, data traceability, SQL audit etc.

Incremental capabilities for traffic control include shadow library, gray release, SQL firewall, blacklist and whitelist, circuit-breaker and rate-limiting, and so on. They are all provided by the database ecosystem layer. However, owing to database fragmentation, it is a huge amount of work to provide full enhancement capability for each database, and there is no unified standard. Database Plus provides users like you with the permutation and combination of supported database types and enhancements by providing a fulcrum.

Pluggable – building a database-oriented functional ecosystem

The Database Plus common layer could become bloated due to docking with an increasing number of database types, and additional enhancement capability. The pluggability borne out of Connect and Enhance is not only the foundation of Database Plus' common layer but also the effective guarantee of infinite ecosystem expansion possibilities.

The pluggable architecture enables Database Plus to truly build a database-oriented functional ecosystem, unifying and managing the global capabilities of heterogeneous databases. It is not only for the distribution of centralized databases but also for the silo function integration of distributed databases.

Microkernel design and pluggable architecture are core values of the Database Plus concept, which is oriented towards a common platform layer rather than a specific function.

Database Plus is the development concept driving ShardingSphere, and ShardingSphere is the best practitioner of the Database Plus concept.

The next sections of this chapter will guide you through the ecosystem's plugin platform that we developed following the Database Plus concept, as well as a few user-defined functions and strategies.

Plugin platform introduction and SPI

Apache ShardingSphere is designed based on a pluggable architecture, where the system provides a variety of plugins that you can choose from to customize your unique system.

The pluggable architecture makes Apache ShardingSphere extremely scalable, allowing you to extend the system based on extension points without changing the core code, making it a developer-oriented design architecture. Developers can easily participate in code development without worrying about the impact on other modules, which also stimulates the open source community and ensures high-quality project development.

The pluggable architecture of Apache ShardingSphere

Apache ShardingSphere is highly scalable through its SPI mechanism, which allows many functional implementations to be loaded into the system. SPI is an API provided by Java, to be implemented or extended by third parties, allowing you to replace and extend Apache ShardingSphere's features in the SPI way. The following diagram illustrates ShardingSphere's pluggable architecture design:

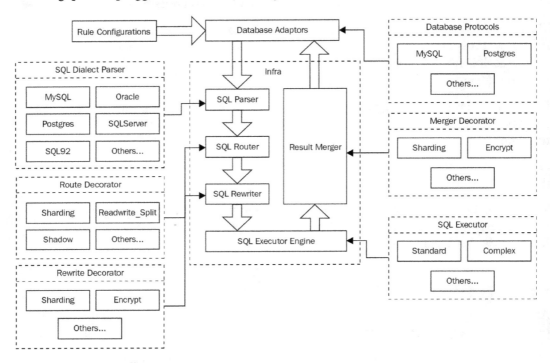

Figure 8.1 – ShardingSphere's pluggable architecture design

As you can see, the pluggable architecture of Apache ShardingSphere consists of a kernel layer, a functional layer, and an ecosystem layer:

- The **kernel layer** is the foundation of the database and contains all the basic capabilities of the database. Based on a plugin-oriented design concept, these functions are abstracted and their implementation can be replaced with pluggable means. Its main capabilities include the query optimizer, distributed transaction engine, distributed execution engine, permission engine, and scheduling engine.

- The **functional layer** mainly provides incremental capabilities for the database. Its main capabilities include data sharding, read/write splitting, database high availability, data encryption, and shadow libraries.

  These features are isolated from each other, and you can choose any combination of features to be used in an overlay. Additionally, you can develop extensions based on the existing extension points, without modifying the kernel code.

- The **ecosystem layer** consists of database protocols, a SQL parser, and a storage adaptor. This layer is used to adapt to and interface with the existing database ecosystem. The database protocol is used to accommodate and serve various dialects of database protocols, the SQL parser is used to interface with various database dialects, and the storage adaptor is used to interface with the database type of the storage node.

These three layers, together with the SPI loading mechanism, constitute a highly scalable and pluggable architecture.

Extensible algorithms and interfaces

This section will provide you with a very useful reference that you will find yourself coming back to quite often in the future.

In the following lists, we will show the SPI for every mode, configuration, kernel, data source, and more. As you can see, for each section (mode, configuration, kernel, and data source) you will be presented with two columns: the left column provides you with the SPI, while the right column provides you with a useful and quick description of what the SPI does:

- **Operation mode**: As you may recall, ShardingSphere operates in three modes – memory mode, standalone mode, and cluster mode. You may choose the one that best fits your scenario's requirements:

  - `StandalonePersistRepository`: Standalone mode configuration information persistence

  - `ClusterPersistRepository`: Cluster mode configuration information persistence

  - `GovernanceWatcher`: Governance monitor

- **Configuration**: ShardingSphere's rule configuration includes multiple SPI interfaces, such as `RuleBuilder` for building rules and `YamlRuleConfigurationSwapper` and `ShardingSphereYamlConstruct` for converting rules:

  - `RuleBuilder`: For converting user configurations into rule objects

  - `YamlRuleConfigurationSwapper`: For converting YAML configurations into standard user configuration

  - `ShardingSphereYamlConstruct`: For converting custom objects to and from YAML

- **Kernel**: The kernel of Apache ShardingSphere offers multiple SPI portals, such as `SQLRouter` and `SQLRewriteContextDecorator`:

  - `SQLRouter`: For processing routing results

  - `SQLRewriteContextDecorator`: For processing SQL rewriting results

  - `SQLExecutionHook`: SQL execution process monitor

  - `ResultProcessEngine`: For processing result sets

  - `StoragePrivilegeHandler`: For handling permission information using database dialects

- **Data sources**: `DataSource` is a data source in ShardingSphere and allows you to manage data sources and the related SPI interfaces for metadata loading:

  - `DatabaseType`: Supported database

  - `DialectTableMetaDataLoader`: For quickly loading metadata using database dialects

  - `DataSourcePoolCreator`: Data source connection pool creator

  - `DataSourcePoolDestroyer`: Data source connection pool destroyer

- **SQL parsing**: ShardingSphere's SQL parsing engine is responsible for parsing different database dialects. You can implement the SPI interface to add new database dialect parsing:

  - `DatabaseTypedSQLParserFacade`: For configuring the lexical and syntactic parser entry for SQL parsing

  - `SQLVisitorFacade`: SQL syntax tree accessor entry

- **Proxy side**: The proxy access end includes the database protocol SPI interface, `DatabaseProtocolFrontendEngine`, and the authority SPI interface, `AuthorityProvideAlgorithm`, for granting permissions:

  - `DatabaseProtocolFrontendEngine`: Protocols for parsing and adapting to accessing the database for ShardingSphere-Proxy

  - `JDBCDriverURLRecognizer`: SQL execution using a JDBC driver

  - `AuthorityProvideAlgorithm`: User rights loading logic

- **Data sharding**: Data sharding is the sharding feature of ShardingSphere. The feature internally provides SPI interfaces such as `ShardingAlgorithm` and `KeyGenerateAlgorithm`:

  - `ShardingAlgorithm`: Sharding algorithm

  - `KeyGenerateAlgorithm`: Distributed primary key generation algorithm

  - `DatetimeService`: Gets the current time for routing

  - `DatabaseSQLEntry`: Gets the database dialect for the current time

- **Read/write splitting**: `ReadWriteSplitting` is the read/write splitting feature of ShardingSphere, which provides SPI interfaces such as `ReplicaLoadBalanceAlgorithm`:

  - `ReadwriteSplittingType`: Read/write splitting type

  - `ReplicaLoadBalanceAlgorithm`: Read library load balancing algorithm

- **High Availability (HA)**: Based on the HA solution, you can achieve high availability of database services by dynamically monitoring changes in the storage nodes:

  - `DatabaseDiscoveryType`: Database discovery types

- **Distributed transactions**: The distributed transaction feature of ShardingSphere provides a solution for data consistency in distributed scenarios, with multiple built-in extensible SPI interfaces:

  - `ShardingSphereTransactionManager`: Distributed transaction manager

  - `XATransactionManagerProvider`: XA distributed transaction manager

  - `XADataSourceDefinition`: Automatically converts non-XA data sources into XA data sources

  - `DataSourcePropertyProvider`: For obtaining the standard properties of data source connection pools

- **Auto scaling**: Scaling supports scaling up and scaling down capabilities in distributed database clusters while considering the user's fast-growing business:

  - `ScalingDataConsistencyCheckAlgorithm`: Data consistency checking algorithm

- **SQL checking**: SQL checking is used to check SQL, and currently implements `AuthorityChecker`:

  - `SQLChecker`: SQL checker

- **Data encryption**: ShardingSphere's data encryption and decryption feature provides the `EncryptAlgorithm` and `QueryAssistedEncryptAlgorithm` SPI interfaces:

  - `EncryptAlgorithm`: Data encryption algorithm

  - `QueryAssistedEncryptAlgorithm`: Query-assisted encryption algorithm

- **Shadow library**: Shadow is the shadow library feature of ShardingSphere, which can meet your needs for online pressure testing. This feature provides the `ShadowAlgorithm` SPI interface:

  - `ShadowAlgorithm`: Shadow library routing algorithm

- **Observability**: Observability is responsible for collecting, storing, and analyzing system observability data, monitoring, and diagnosing system performance. By default, it provides support for SkyWalking, Zipkin, Jaeger, and OpenTelemetry:

  - `PluginDefinitionService`: Agent plugin definition

  - `PluginBootService`: Plugin launch service definition

This completes the SPI configurations. Now, let's look at the functions and strategies that you can customize by yourself.

User-defined functions and strategies – SQL parser, sharding, read/write splitting, distributed transactions

In this section, you will learn how to create and configure custom functions. We will start with the SQL parser before looking at data sharding, read/write splitting, and, finally, distributed transactions. The examples and steps you will find in the following sections will empower you to make Apache ShardingSphere truly yours.

Customizing your SQL parser

This section will describe how to use the SQL parser engine, which is compatible with different database dialects. By parsing SQL statements to a message that can be understood by Apache ShardingSphere, enhanced database features can be achieved. Various dialect parsers that are part of the SQL parser are loaded through the SPI method. Therefore, you can handily develop or enrich database dialects.

Implementation

Let's review the necessary code to get started. In this section, you will learn how to parse SQL into a statement and how to format SQL.

First, let's parse SQL into a statement:

```Java
CacheOption cacheOption = new CacheOption(128, 1024L, 4);
SQLParserEngine parserEngine = new SQLParserEngine("MySQL",
cacheOption, false);
ParseContext parseContext = parserEngine.parse("SELECT t.id,
t.name, t.age FROM table1 AS t ORDER BY t.id DESC;", false);
SQLVisitorEngine visitorEngine = new SQLVisitorEngine("MySQL",
"STATEMENT", new Properties());
MySQLStatement sqlStatement = visitorEngine.
visit(parseContext);
System.out.println(sqlStatement.toString());
```

Next, let's format SQL:

```Java
Properties props = new Properties();
props.setProperty("parameterized", "false");
CacheOption cacheOption = new CacheOption(128, 1024L, 4);
SQLParserEngine parserEngine = new SQLParserEngine("MySQL",
cacheOption, false);
ParseContext parseContext = parserEngine.parse("SELECT age AS
b, name AS n FROM table1 JOIN table2 WHERE id = 1 AND name =
'lu';", false);
SQLVisitorEngine visitorEngine = new SQLVisitorEngine("MySQL",
"FORMAT", props);
String result = visitorEngine.visit(parseContext);
System.out.println(result);
```

Now, let's review the extensible algorithms at your disposal.

Extensible algorithms

As we mentioned previously, we are going to provide you with extensible algorithms. This will be the format we will follow throughout this chapter – first, you will review the implementation, and then move on to the extensible algorithms for each feature.

ShardingSphere provides the following extensible access for its parsing engine, making it convenient to parse other SQL dialects:

- Configure lexical and syntactic parsers for the SQL parser interface:

```css
org.apache.shardingsphere.sql.parser.spi.
DatabaseTypedSQLParserFacade
```

- Configure the SQL syntax tree visitor interface:

```css
org.apache.shardingsphere.sql.parser.spi.SQLVisitorFacade
```

The next section will introduce you to the extensible algorithms of ShardingSphere's original feature: data sharding.

Customizing the data sharding feature

The sharding function can horizontally divide user data, improve system performance and usability, and slash operation and maintenance costs. The sharding function also provides rich extensibility. You can expand the sharding algorithm and distributed sequence algorithm based on the SPI mechanism.

The following sections will show you how to implement a sharding algorithm by following a simple two-step process.

Implementation

Let's begin:

1. STANDARD, COMPLEX, and HINT are supported for implementing their corresponding algorithms. Here, we will use STANDARD as an example:

```
public class MyDBRangeShardingAlgorithm
implementsStandardShardingAlgorithm<Integer> {
```

```
        @Override
        public String doSharding(final Collection<String>
availableTargetNames, finalPreciseShardingValue<Integer>
shardingValue) {
                return null;
        }
        @Override
        public String getType() {
                return "CLASS_BASED";
        }
```

2. Add SPIconfiguration, as follows (please refer to the Java standard SPI loading method):

```css
org.apache.shardingsphere.sharding.spi.ShardingAlgorithm
```

3. Next, let's apply the algorithm to implement a user-defined distributed sequence algorithm:

```typescript
public final class IncrementKeyGenerateAlgorithm
implements KeyGenerateAlgorithm {

        private final AtomicInteger count = new
AtomicInteger();

        @Override
        public Comparable<?> generateKey() {
                return count.incrementAndGet();
        }

        @Override
        public String getType() {
                return "INCREMENT";
        }
}
```

4. Add the SPI configuration (please refer to the Java standard SPI loading method):

```
org.apache.shardingsphere.sharding.spi.
KeyGenerateAlgorithm
```

Extensible algorithms

ShardingSphere provides you with extensible access to user-defined sharding algorithms and distributed sequence algorithms.

Use the following code to gain user-defined sharding algorithm access:

```
org.apache.shardingsphere.sharding.spi.ShardingAlgorithm
```

Use the following code to gain user-defined distributed sequence algorithm access:

```
org.apache.shardingsphere.sharding.spi.KeyGenerateAlgorithm
```

The next section will guide you through the read/write splitting implementation and its extensible algorithms.

Read/write splitting

Read/writing splitting refers to a splitting architecture that divides a database into a primary database and a secondary database. The primary database is in charge of adding, deleting, and modifying transactions, while the secondary database queries operations.

Read/write splitting can effectively improve the throughput and usability of the system. Apache ShardingSphere's read/write splitting functionality provides an extensible SPI for the read database load balancing algorithm. This section will mainly discuss how to use the user-defined load balancing algorithm.

Implementation

The following code shows how to access the read database load balancing algorithm:

```TypeScript

@Getter
@Setter
public final class TestReplicaLoadBalanceAlgorithm implements
ReplicaLoadBalanceAlgorithm {
```

```
        private Properties props = new Properties();

        @Override
        public String getDataSource(final String name,
final String writeDataSourceName, final List<String>
readDataSourceNames) {
                return null;
        }

        @Override
        public String getType() {
                return "TEST";
        }
}
```

You can add the SPI configuration as follows (please refer to the Java standard SPI loading method):

```
org.apache.shardingsphere.readwritesplitting.spi.
ReplicaLoadBalanceAlgorithm
```

Extensible algorithms

The user-defined load balancing algorithm can be extended through the following access:

```
org.apache.shardingsphere.readwritesplitting.spi.
ReplicaLoadBalanceAlgorithm
```

In the next section, you will learn how to customize distributed transactions with your strategy.

Distributed transactions

Distributed transactions ensure that the transaction semantics of multiple storage databases are managed by ShardingSphere. Choosing ShardingSphere will feel just like using a traditional database, with common functions such as `begin`, `commit`, `rollback`, and `set autocommit`.

Implementation

You should start by configuring the `server.yaml` transaction configuration according to the following code example:

```YAML
-- XA provides consistent semantics
rules:
  - !TRANSACTION
    defaultType: XA
    providerType: Narayana
```

Once you have configured the `server.yaml` transaction configuration, you can configure the table's structure and its initial data:

```SQL
create table account (
    id int primary key,
    balance int
);


insert into account values(1,0),(2,100);
```

If you're using Java code, please refer to this option for low-level MySQL. The configuration for PostgreSQL is the same except that `jdbcurl` needs to be modified:

```
public void test() {
        Connection connection = null;
        try {
            connection = DriverManager.
getConnection("jdbc:mysql://127.0.0.1:3307/test", "root",
"root");
            connection.setAutoCommit(false);
            ......
            connection.commit();
        } catch (Exception e) {
            connection.rollback();
        } finally {
            ......
```

```
                    }
              }
```

If you're using a MySQL client, please refer to this option for low-level MySQL. The configuration for PostgreSQL is the same:

```
-- Start transaction and commit
mysql> BEGIN;
mysql> UPDATE account SET balance=1 WHERE id=1;
mysql> SELECT * FROM account;
mysql> UPDATE account SET balance=99 WHERE id=2;
mysql> SELECT * FROM account;
mysql> COMMIT;
mysql> SELECT * FROM account;

-- Start transaction and rollback
mysql> BEGIN;
mysql> SELECT * FROM account;
mysql> UPDATE account SET balance=0 WHERE id=1;
mysql> SELECT * FROM account;
mysql> UPDATE account SET balance=100 WHERE id=2;
mysql> SELECT * FROM account;
mysql> ROLLBACK;
mysql> SELECT * FROM account;
```

Extensible algorithms

Distributed transactions provide integration with Narayana, Atomikos, and Base. If you are looking to customize TM, you can achieve expansion by providing the following access:

```
org.apache.shardingsphere.transaction.spi.
ShardingSphereTransactionManager
```

Now, let's learn how to implement and configure functions and strategies.

User-defined functions and strategies – encryption, SQL authority, user authentication, shadow DB, distributed governance

In the following sections, you will learn how to configure data encryption, SQL parsing, SQL authority, user authentication, shadow DB, and distributed governance. We will follow the same format we used previously – that is, starting with the implementation, then looking at the extensible algorithms.

Data encryption

Data encryption is a way to ensure data security by data transformation. The Apache ShardingSphere data encryption feature provides significant extensibility, allowing you to extend the data encryption algorithm and query-assisted column data encryption algorithm based on the SPI mechanism.

Implementing your encryption algorithm

Let's begin:

1. You can implement the EncryptAlgorithm interface using the following code:

```
public final class NormalEncryptAlgorithmFixture
implements EncryptAlgorithm<Object, String> {

    @Override
    public String encrypt(final Object plainValue) {
        return "encrypt_" + plainValue;
    }

    @Override
    public Object decrypt(final String cipherValue) {
        return cipherValue.replaceAll("encrypt_", "");
    }
```

2. Add the SPI configuration (please refer to the Java standard SPI loading method):

```
org.apache.shardingsphere.encrypt.spi.EncryptAlgorithm
```

3. You can implement the user-defined QueryAssistedEncryptAlgorithm interface for column data as per the following code:

```
public final class QueryAssistedEncryptAlgorithmFixture
implementsQueryAssistedEncryptAlgorithm<Object, String> {


    @Override
    public String queryAssistedEncrypt(final Object
plainValue) {
            return "assisted_query_" + plainValue;
    }
```

4. Add the SPI configuration (please refer to the Java standard SPI loading method).

Since QueryAssistedEncryptAlgorithm inherited the
EncryptAlgorithm access, the SPI file is named after it – that is,
EncryptAlgorithm:

```
org.apache.shardingsphere.encrypt.spi.EncryptAlgorithm
```

Extensible algorithms

ShardingSphere provides you with user-defined extensible access to encryption and decryption algorithms:

```
org.apache.shardingsphere.encrypt.spi.EncryptAlgorithm
```

In the next section, you will learn how to customize the user authentication feature.

User authentication

As we mentioned in *Chapter 4, Key Features and Use Cases – Focusing on Performance and Security*, Apache ShardingSphere supports handshakes and authentication for various database connection protocols through its Authentication Engine.

Implementation

Generally, you won't need to be concerned with ShardingSphere's internal implementation. Similar to using MySQL and PostgreSQL, you only have to create a connection with ShardingSphere via a terminal or visualization client. Once done, you only have to enter or configure the proper username and password.

On the other hand, if you'd like to configure a new database authentication type or use a customized authentication method (such as a user-defined password verification algorithm), you should know how to extend Authentication Engine. The next section will show you how to do that.

Extensible algorithms

AuthenticationEngine is an adaptor. Its definition is as follows:

```Java
public interface AuthenticationEngine {

    /**
     * Handshake.
     */
    int handshake(ChannelHandlerContext context);

    /**
     * Authenticate.
     */
    AuthenticationResult authenticate(ChannelHandlerContext
context, PacketPayload payload);
}
```

Judging from its definition, we can tell that the AuthenticationEngine access is very neat. It follows these two methods:

- A handshake is used to respond to a handshake demand from the client's end.
- authenticate is responsible for verifying the client's username and password.

At the time of writing, Apache ShardingSphere provides three ways of implementing `AuthenticationEngine`. The following list specifies the name and compatible database type:

- `MySQLAuthenticationEngine`: MySQL
- `PostgreSQLAuthenticationEngine`: PostgreSQL
- `OpenGaussAuthenticationEngine`: OpenGauss

If you need ShardingSphere to be compatible with a new type of database, such as Oracle, you can add another `AuthenticationEngine` access. If you're only hoping to adjust user password verification rules, you can customize `authenticate` access as per your needs.

In the future, as ShardingSphere supports more databases and verification methods, the implementation of `AuthenticationEngine` will be expanded, and its functions will be improved.

Remaining on the theme of database security, in the next section, you will learn how to customize the SQL authority feature.

SQL authority

When we refer to SQL authority, we mean that after receiving SQL commands entered by users, Apache ShardingSphere will check whether they have related authorities according to the operation types and data scopes of their commands, and permit or decline said operations.

For more concepts and background information about SQL authority, please refer to *Chapter 4, Key Features and Use Cases – Focusing on Performance and Security*.

Implementation

In ShardingSphere, you can choose different types of authority providers to achieve authority controls at different levels. The authority provider needs to be configured in `server.yaml`. The format is as follows:

```YAML
rules:
  - !AUTHORITY
    users:
      - root@%:root
      - sharding@:sharding
```

```
provider:
    type: ALL_PRIVILEGES_PERMITTED
```

Here, the input options are ALL_PRIVILEGES_PERMITTED and SCHEMA_PRIVILEGES_PERMITTED:

- ALL_PRIVILEGES_PERMITTED: Using this provider means that the user has access to all privileges. This means that in ShardingSphere, the user can perform all operations on all tables and databases. When no provider type is designated, this provider will be used by default.

- SCHEMA_PRIVILEGES_PERMITTED: This refers to controlling user authority at the schema level. For detailed usage, please refer to the introduction in *Chapter 4, Key Features and Use Cases – Focusing on Performance and Security*.

Extensible algorithms

From the previous section, we know that Apache ShardingSphere achieves authority controls at different levels by using different authority providers. All authority providers achieve SPI access via AuthorityProvideAlgorithm. Its definition is as follows:

```Java
public interface AuthorityProvideAlgorithm extends
ShardingSphereAlgorithm {

    /**
     * Initialize authority.
     */
    void init(Map<String, ShardingSphereMetaData> metaDataMap,
Collection<ShardingSphereUser> users);

    /**
     * Refresh authority.
     */
    void refresh(Map<String, ShardingSphereMetaData>
metaDataMap, Collection<ShardingSphereUser> users);

    /**
     * Find Privileges.
     */
```

```
      Optional<ShardingSpherePrivileges> findPrivileges(Grantee
grantee);
}
```

Let's learn more about the terms used in the preceding code:

- `init` is used to initialize the provider – for instance, parsing the configuration of props into the necessary format.

- `refresh` is for refreshing information that will be used when you're dynamically updating the user and its authority.

- `findPrivileges` is for finding the privilege list of target users. It will return to a collection of `ShardingSpherePrivileges`.

How does ShardingSphere determine a user's privileges on a specific subject? It is determined by `ShardingSpherePrivileges` access. The `ShardingSpherePrivileges` access definition is as follows:

- Java

```java
public interface ShardingSpherePrivileges {

    /**
     * Set super privilege.
     */
    void setSuperPrivilege();

    /**
     * Has privileges of schema.
     */
    boolean hasPrivileges(String schema);

    /**
     * Has specified privileges.
     */
    boolean hasPrivileges(Collection<PrivilegeType>
privileges);

    /**
     * Has specified privileges of subject.
```

```
    */
    boolean hasPrivileges(AccessSubject accessSubject,
Collection<PrivilegeType> privileges);
}
```

`ShardingSpherePrivileges` access has four methods, as follows:

- The `setSuperPrivilege` method is used to set super privileges for users.
- The other three `hasPrivilege` methods are used to check whether a user has the privilege of a specific operation or object.

In this section, you learned that the access ShardingSphere provides for SQL authority is standard and neat. If you want a customized authority method, you can make one by extending the access of `AuthorityProvideAlgorithm` and `ShardingSpherePrivileges`.

In the next section, you will learn how to customize the shadow DB feature according to your preference.

Shadow DB

Apache ShardingSphere's shadow DB is quite extensible: based on the SPI mechanism being used, it supports the extending column shadow algorithm and the HINT shadow algorithm.

Implementation

Configuring the shadow DB feature will come very easily to you, thanks to ShardingSphere's user-friendliness. Follow these steps:

1. Let's start by configuring a user-defined `ColumnShadowAlgorithm`:

```Java
public class CustomizeColumnMatchShadowAlgorithm
implements ColumnShadowAlgorithm<Comparable<?>> {

    @Override
    public void init() {
    }

    @Override
```

```
        public boolean isShadow(final
PreciseColumnShadowValue<Comparable<?>>
preciseColumnShadowValue) {
                // TODO Custom Shadow Algorithm Judgment
                return true/false;
        }


        @Override
        public String getType() {
                return "CUSTOMIZE_COLUMN";
        }
}
```

The default method is set up as `isShadow`: `PreciseColumnShadowValue`: `PreciseColumnShadowValue`:

```Java
public final class PreciseColumnShadowValue<T extends
Comparable<?>> implements ShadowValue {

        private final String logicTableName;

        private final ShadowOperationType
shadowOperationType;

        private final String columnName;

        private final T value
```

2. Next, let's configure the user-defined `HintShadowAlgorithm`:

```Java
public final class CustomizeHintShadowAlgorithm
implements HintShadowAlgorithm<String> {

        @Override
        public void init() {
        }
```

```
        @Override
        public boolean isShadow(final Collection<String>
relatedShadowTables, final PreciseHintShadowValue<String>
preciseHintShadowValue) {
                // TODO Custom Shadow Algorithm Judgment
                return true/false;
        }


        @Override
        public String getType() {
                return "CUSTOMIZE_HINT";
        }
}
```

Once you have included a custom strategy, as we did in *Step 2*, if you query the system, you will be able to see the difference in the Shadow strategy, as follows:

- relatedShadowTables: Shadow tables configured in configuration files
- preciseHintShadowValue: A precise HINT shadow value

Let's query the code to see this difference:

```Java
public final class PreciseHintShadowValue<T extends
Comparable<?>> implements ShadowValue {

        private final String logicTableName;

        private final ShadowOperationType
shadowOperationType;

        private final T value;
}
```

As we can see, the HINT shadow algorithm's value type is String.

3. Add the SPI configuration (please refer to the standard loading method of the **Java Service Provider Interface (SPI)**):

    ```
    org.apache.shardingsphere.shadow.spi.ShadowAlgorithm
    ```

Extensible algorithms

At the time of writing, Apache ShardingSphere provides the `Shadow Algorithm` SPI with three different implementations:

- `ColumnValueMatchShadowAlgorithm`: Shadow algorithm matching column value

- `ColumnRegexMatchShadowAlgorithm`: Shadow algorithm matching column regex

- `SimpleHintShadowAlgorithm`: Simple `HINT` shadow algorithm

Now that you have a reference, you can come back to this section when you're customizing your shadow DB feature. Now, let's learn how to customize the distributed governance feature.

Distributed governance

Apache ShardingSphere leverages the third-party `Zookeeper` and `etcd` components to implement distributed governance. It can flexibly integrate with other components to meet your needs for different stacks or in different scenarios. This section will show you how to integrate with other components to extend ShardingSphere's distributed governance capability.

Implementation

Distributed governance is available in ShardingSphere's cluster mode. You need to configure the cluster mode and define a distributed storage strategy in `server.yaml` to use the user-defined distributed governance function. The configuration method is shown here:

```yaml
YAML
mode:
  type: Cluster
  repository:
    type: CustomRepository # custom repository type
    props: # custom properties
       custom-time-out-: 30
       custom-max-retries: 5
  overwrite: false
```

To enable the aforementioned configuration in ShardingSphere, follow these steps:

1. Implement the Distributed Persist Strategy SPI and define the implementation type as `CustomRepository`:

    ```Java
    org.apache.shardingsphere.mode.repository.cluster.
    ClusterPersistRepository
    ```

    The following is a referential implementation class:

    ```Java
    public final class CustomPersistRepository implements
    ClusterPersistRepository{

        @Override
        public String get(final String key) {
            return null;
        }
        @Override
        public List<String> getChildrenKeys(final String
    key) {
            return null;
        }
        @Override
        public void persist(final String key, final String
    value) {
        }
        @Override
        public void delete(final String key) {
        }
        @Override
        public void close() {
        }
        @Override
        public void init(final
    ClusterPersistRepositoryConfiguration config) {
        }
        @Override
    ```

```java
    public void persistEphemeral(final String key, final
String value) {
    }

    @Override
    public String getSequentialId(final String key,
final String value) {
        return null;
    }

    @Override
    public void watch(final String key, final
DataChangedEventListener listener) {
    }

    @Override
    public boolean tryLock(final String key, final long
time, final TimeUnit unit) {
        return false;
    }

    @Override
    public void releaseLock(final String key) {
    }

    @Override
    public String getType() {
        return "CustomRepository";
    }
}
```

2. Add the SPI configuration (please refer to the standard loading method of the Java Service Provider Interface).

3. To parse the following user-defined **configuration items (CIs)**, you need to implement the SPI for mapping CIs:

```yaml
YAML
props: # custom properties
        custom-time-out: 30
        custom-max-retries: 5
```

The following code shows the CI mapping's SPI abstract class:

```Java
org.apache.shardingsphere.infra.properties.
TypedPropertyKey
```

The following code shows a referential implementation class:

```Java
@RequiredArgsConstructor
@Getter
public enum CustomPropertyKey implements TypedPropertyKey
{

    CUSTOM_TIME_OUT("customTimeOut", String.valueOf(30),
long.class),

    CUSTOM_MAX_RETRIES("customMaxRetries", String.
valueOf(3), int.class)
    ;

    private final String key;

    private final String defaultValue;

    private final Class<?> type;
}
```

4. Transforming the Inherit-config property of the SPI abstract class is used to get the user-defined configurations in YAML:

```Java
org.apache.shardingsphere.infra.properties.
TypedProperties
```

The following code shows a referential implementation class:

```Scala
public final class CustomProperties extends
TypedProperties<CustomPropertyKey> {
```

```
public CustomProperties(final Properties props) {
        super(CustomPropertyKey.class, props);
    }
}
```

Among the user-defined distributed storage strategy implementation classes, you can use the `init` method to get the user-defined configurations and initialize them:

```Java
@Override
public void init(final
ClusterPersistRepositoryConfiguration config) {
    CustomProperties properties = new
CustomProperties(config.getProps());
    long customTimeOut = properties.
getValue(CustomPropertyKey.CUSTOM_TIME_OUT);
    long customMaxRetries = properties.
getValue(CustomPropertyKey.CUSTOM_MAX_RETRIES);
}
```

Extensible algorithms

The following code shows the user-defined configuration item mapping interface:

```Java
org.apache.shardingsphere.infra.properties.TypedPropertyKey
```

The following code shows the user-defined configuration item parsing interface:

```Java
org.apache.shardingsphere.infra.properties.TypedProperties
```

The following code shows the user-defined distributed persist strategy interface:

```Java
org.apache.shardingsphere.mode.repository.cluster.
ClusterPersistRepository
```

In the next section, you will learn how to customize the scaling feature.

Scaling

As per the design concept of Apache ShardingSphere, scaling also provides rich extensibility, allowing you to configure most of its parts. If there is an existing implementation that fails to meet your requirements, you can leverage SPI to make extensions by yourself, such as the data consistency checker.

Implementing the data consistency checker algorithm

The main SPI algorithm that we use is `DataConsistencyCheckAlgorithm`.

Generally, most methods provide algorithm metadata information. When you leverage DistSQL to manually trigger the data consistency checker, you can specify the algorithm's type (concurrently, there is a DistSQL statement that lists all algorithms).

The core `getSingleTableDataCalculator` method is based on the algorithm's type and database's type and obtains the right single table data of `SingleTableDataCalculator` to calculate the SPI implementation:

```Java
public interface DataConsistencyCheckAlgorithm extends
ShardingSphereAlgorithm, ShardingSphereAlgorithmPostProcessor,
SingletonSPI {

    /**
     * Get algorithm description.
     *
     * @return algorithm description
     */
    String getDescription();

    /**
     * Get supported database types.
     *
     * @return supported database types
     */
    Collection<String> getSupportedDatabaseTypes();

    /**
     * Get algorithm provider.
```

```
 *
 * @return algorithm provider
 */
String getProvider();

/**
 * Get single table data calculator.
 *
 * @param supportedDatabaseType supported database type
 * @return single table data calculator
 */
SingleTableDataCalculator
getSingleTableDataCalculator(String supportedDatabaseType);
}
```

The next sub-algorithm SPI we will look at is SingleTableDataCalculator.

The SingleTableDataCalculator algorithm provides single table data calculation capabilities for the main DataConsistencyCheckAlgorithm algorithm SPI. check and checksum are usually used for calculation.

Based on different *algorithm types* and *database types*, the implementation of this interface supports heterogeneous database migration, and can perform calculations on the source and target ends separately:

```Java
public interface SingleTableDataCalculator {

    /**
     * Get algorithm type.
     *
     * @return algorithm type
     */
    String getAlgorithmType();

    /**
     * Get database type.
     *
     * @return database type
```

```
    */
    String getDatabaseType();

    /**
     * Calculate table data, usually checksum.
     *
     * @param dataSourceConfig data source configuration
     * @param logicTableName logic table name
     * @param columnNames column names
     * @return calculated result, it will be used to check
equality.
     */
    Object dataCalculate(ScalingDataSourceConfiguration
dataSourceConfig, String logicTableName, Collection<String>
columnNames);
}
```

Extensible algorithms

Now, let's look at the extensible algorithms. The following list provides you with a useful and quick description of what the algorithm does:

- For the `ScalingDataConsistencyCheckAlgorithm` algorithm, we have the `ScalingDefaultDataConsistencyCheckAlgorithm` implementation class, a consistency check algorithm based on CRC32 matching.

- For the `SingleTableDataCalculator` algorithm, we have the `DefaultMySQLSingleTableDataCalculator` implementation class, a single table data calculation algorithm based on CRC32 matching (only applicable for MySQL).

Now that we've looked at feature customization, let's learn how to fine-tune ShardingSphere-Proxy's properties.

ShardingSphere-Proxy – tuning properties and user scenarios

In this section, we will look at the property parameters of ShardingSphere-Proxy. In the props section of the `server.yaml` configuration file, some parameters are related to functions, while some parameters are related to performance.

Adjusting performance-oriented parameters in specific scenarios can improve the performance of ShardingSphere-Proxy as much as possible when environment resources are limited. Next, we will describe the parameters of props.

Properties introduction

Let's review the performance-oriented parameters of ShardingSphere-Proxy. In this section, you will find each of these performance-oriented parameters, along with a brief description of what they do.

max-connections-size-per-query

The default value of the `max-connections-size-per-query` parameter is 1.

When using ShardingSphere to execute a line of SQL, this parameter mainly controls the maximum connections that are allowed to be fetched from each data source. In the data sharding scenario, when a line of logic SQL is routed to multiple shards, increasing the parameter can raise the concurrency of actual SQL and reduce the time consumption of the query.

kernel-executor-size

The default value of the `kernel-executor-size` parameter is 0.

This parameter mainly controls the internal part of ShardingSphere and the size of thread pools that are used to execute SQLs. The default value is 0. Using `java.util.concurrent.Executors#newCachedThreadPool` means that there's no limit for threads. This parameter is mainly used in data sharding.

Normally, without special adjustment, the thread pool will create or delete threads according to your needs. You can set a fixed value to reduce the consumption of thread creation or limit resource consumption.

sql-show

The default value of the `sql-show` parameter is `false`.

When this parameter is on, SQL (actual SQL) that's been processed by the original SQL (logic SQL) and the kernel will be output to a log. Since log output has a significant influence on performance, it is advised to only enable this when necessary.

sql-simple

The default value of the `sql-simple` parameter is `false`.

It only works when `sql-show` is set to `true`. If the parameter is set to `true`, the log will not output detailed parameters regarding the placeholders in `Prepared Statement`.

show-process-list-enabled

The default value of the `show-process-list-enabled` parameter is `false`.

This parameter controls whether the `showprocesslist` function is enabled. It only works in Cluster mode. This function is similar to MySQL's `showprocesslist`. At the time of writing, it only works for DDL and DML statements.

check-table-metadata-enabled

The default value of the `check-table-metadata-enabled` parameter is `false`.

It allows a sharding table metadata information check to be performed. If the parameter is `true`, the metadata of all the tables will be loaded when the sharding table metadata is loaded, and their consistency will be checked.

check-duplicate-table-enabled

The default value of the `check-duplicate-table-enabled` parameter is `false`.

It checks duplicate tables. If the parameter is `true`, when initializing a single table, the existence of duplicate tables will be checked. If there is a duplicate table, the output will be abnormal.

sql-federation-enabled

The default value of the `sql-federation-enabled` parameter is `false`.

It enables the SQL Federation execution engine. If the parameter is `true`, SQL Federation can be used to execute an engine that supports inter-database distributed queries. At the time of writing, when the table data amount is large, using the Federation execution engine may cause ShardingSphere-Proxy to consume more CPU and memory.

proxy-opentracing-enabled

The default value of the `proxy-opentracing-enabled` parameter is `false`. This parameter enables or disables OpenTracing-related functions.

proxy-hint-enabled

The default value of the `proxy-hint-enabled` parameter is `false`.

This parameter controls whether the `hint` function of ShardingSphere-Proxy should be enabled. When the `hint` function is enabled, requests that have been sent to ShardingSphere-Proxy MySQL will be processed by the exclusive thread of each client end, which may compromise ShardingSphere-Proxy's performance.

proxy-frontend-flush-threshold

The default value of the `proxy-frontend-flush-threshold` parameter is `128`.

This parameter mainly controls the frequency of the `flush` operation in the buffer zone when ShardingSphere-Proxy sends query results to the client's end. For instance, if there are 1,000 lines of query results and the flush threshold is set to 100, ShardingSphere-Proxy will perform one `flush` operation every 100 lines of data it sends to the client end.

Properly reducing the parameter may make it faster for the client end to receive response data and a lower SQL response time. However, frequent `flush` operations may increase the network load.

proxy-backend-query-fetch-size

The default value of the `proxy-backend-query-fetch-size` parameter is `-1`.

It only works when `ConnectionMode` is `Memory Strictly`. When ShardingSphere-Proxy uses JDBC to execute select statements in databases, this parameter can control the minimum lines of data that ShardingSphere-Proxy fetches. It is similar to cursor. The default value of this parameter is `-1`, which means minimizing the fetch size.

Setting a relatively small fetch size may reduce the memory occupation of ShardingSphere-Proxy. However, it may increase the interactions between ShardingSphere-Proxy and the database, resulting in a longer SQL response and vice versa.

proxy-frontend-executor-size

The default value of the `proxy-frontend-executor-size` parameter is `0`.

ShardingSphere-Proxy uses **Netty** to implement database protocols and communication with the client end. This parameter mainly controls the thread size of `EventLoopGroup`, which is used by ShardingSphere-Proxy. Its default value is `0`, which means that Netty determines the thread size of `EventLoopGroup`. Normally, it is twice the size of the available CPU.

proxy-backend-executor-suitable

The available options for the `proxy-backend-executor-suitable` parameter are OLAP (default) and OLTP.

When the number of client ends that connect with ShardingSphere-Proxy is rather small (less than `proxy-frontend-executor-size`) and the client ends are not required to execute time-consuming SQL, using OLTP may reduce SQL execution time consumption at the ShardingSphere-Proxy layer. If you are unsure whether using OLTP can improve performance, you're advised to use the default value – that is, OLAP.

This configuration influences ShardingSphere-Proxy when it's used to execute SQL's thread pool choice. When using OLAP, ShardingSphere-Proxy will use an independent thread pool to run logic, such as SQL routing, rewriting, interacting with databases, and certain thread switches that may exist.

If OLTP is being used, ShardingSphere-Proxy will directly use Netty's `EventLoop` thread, and the logic that's been requested to be processed on one client end, such as SQL routing, rewriting, and interaction with databases, will be executed in one thread. Compared to OLAP, this may reduce SQL time consumption at the ShardingSphere-Proxy layer.

OLTP can also be used since the inherently synchronous JDBC is used when ShardingSphere-Proxy interacts with a database. If Netty's `EventLoop` were accidentally obstructed due to slow SQL execution or lengthened interaction between ShardingSphere-Proxy and the database for other reasons, the response of other client ends connected to ShardingSphere-Proxy would be extended.

proxy-frontend-max-connections

The default value of the `proxy-frontend-max-connections` parameter is 0.

This configuration controls the minimum TCP connections that client ends are allowed to create with ShardingSphere-Proxy. It is similar to MySQL's `max_connections` configuration. The default value is 0. A value less than or equal to 0 means no limit on connections.

Extensible algorithms

We will start this section by covering some testing scenarios. Here, we will showcase how to adjust ShardingSphere-Proxy's attribute parameters in specific scenarios to maximize ShardingSphere-Proxy's resource usage and improve its performance in case of limited resources.

Case 1 – Using ShardingSphere-Proxy's PostgreSQL to test performance with a sharding rule in BenchmarkSQL 5.0

In this section, we will use ShardingSphere-Proxy's PostgreSQL to perform a TPC-C benchmark test while implementing the data sharding scenario. Here, you will learn how to adjust the parameters of ShardingSphere-Proxy and the client to greatly improve test results.

Requirement: In the data sharding scenario, each logic SQL needs to match its unique actual SQL.

In the BenchmarkSQL 5.0 `PostgreSQL JDBC URL` configuration file, you need to configure the following two parameters:

- The following is the first parameter we will look into:

  ```
  defaultRowFetchSize=100
  ```

  The suggested value is `100` but you should still adjust it based on the actual test situation. `BenchmarkSQL` only fetches one row of data from `ResultSet` in most queries, and sometimes fetches multiple rows in a few queries.

  By default, PostgreSQL JDBC may fetch 1,000 rows of data from proxy every time, but the actual test logic only requires a small amount of data. It's necessary to adjust the `defaultRowFetchSize` value of PostgreSQL JDBC to reduce its interactions with ShardingSphere-Proxy as much as possible on the premise that the `BenchmarkSQL` logic is still satisfied.

- The following is the second parameter that we will look into:

  ```
  reWriteBatchedInserts=true
  ```

  By rewriting the `Insert` statement to combine multiple sets of parameters into the values of the same `Insert` statement, this option can reduce interactions between `BenchmarkSQL` and ShardingSphere-Proxy when you're initializing data and the `New Order` business. Then, it can reduce network transmission time that, in most cases, takes a relatively high proportion of total SQL execution time.

  The following code can be used as a reference for configuring `BenchmarkSQL 5.0`:

  ```
  Makefile
  db=postgres
  driver=org.postgresql.Driver
  conn=jdbc:postgresql://localhost:5432/postgres?defaultRow
  FetchSize=100&reWriteBatchedInserts=true
  ```

Configuring `server.yaml` is recommended. In terms of other parameters, you can use the default values or adjust them:

```yaml
YAML
props:
  proxy-backend-query-fetch-size: 1000
  # According to the actual stress test results,
concurrently enable or remove the following two
parameters
  proxy-backend-executor-suitable: OLTP # Disable it
during data initialization and enable it during stress
tests
  proxy-frontend-executor-size: 200 # Keep consistent
with Terminals
```

- Next, we have the `proxy-backend-query-fetch-size` parameter.

The following is a SQL statement that queries the `Delivery` property of BenchmarkSQL:

```sql
SQL
SELECT no_o_id
      FROM bmsql_new_order
      WHERE no_w_id = ? AND no_d_id = ?
      ORDER BY no_o_id ASC
```

In terms of the query, its results may exceed 1,000 rows. `BenchmarkSQL` gets query results with a fetch size of `1000` from ShardingSphere-Proxy. ShardingSphere-Proxy's `ConnectionMode` is `Memory Strictly` and `proxy-backend-query-fetch-size` is set as the default value (this amounts to a fetch size of 1). This means that every time `BenchmarkSQL` executes one SQL query, ShardingSphere-Proxy needs to fetch more than 1,000 times from the database, which results in a long SQL execution time and lower performance test results.

- Now, let's look at the `proxy-backend-executor-suitable` and `proxy-frontend-executor-size` parameters.

In the BenchmarkSQL 5.0 testing scenarios, there is no long-time SQL execution, so users can try to enable the OLTP option of `proxy-backend-executor-suitable` to reduce the cross-thread overhead of frontend and backend logic.

You should use the OLTP option and the `proxy-frontend-executor-size` option together. It is suggested to set the parameter value to `Terminals`.

Case 2 – ShardingSphere kernel parameter tuning scenario

ShardingSphere's kernel-related adjustable parameters include `check-table-metadata-enabled`, `kernel-executor-size`, and `max-connections-size-per-query`. Let's look at their usage scenarios and provide some useful tips:

- The `check-table-metadata-enabled` parameter is used to enable/disable sharding table metadata information verification. If its value is `true`, when you load the metadata of the sharding table, the metadata of all the tables is loaded, and a metadata consistency check is performed. The metadata checker needs different amounts of time to complete consistency verification and its duration is dependent on the number of actual tables. If the user can ensure that the structure of all actual tables is consistent, then they can set this parameter to `false` to speed up ShardingSphere's bootup time.

- The `kernel-executor-size` parameter is used to control the internal part of ShardingSphere and the size of the thread pools that are used to execute SQLs. To set the parameter, users need to consider the amount of SQL to be executed in the data sharding scenario and choose a fixed value that covers multiple scenarios. This will reduce the overhead of thread creation and restrict resource usage.

- The `max-connections-size-per-query` parameter manages the maximum number of connections that are allowed for each data source when ShardingSphere is used to execute a SQL statement. Taking the following `t_order` table configuration as an example, let's look at the functionality of `max-connections-size-per-query`:

```
YAML
rules:
- !SHARDING
  tables:
    t_order:
      actualDataNodes: ds_${0..1}.t_order_${0..1}
      databaseStrategy:
        standard:
          shardingColumn: user_id
          shardingAlgorithmName: database_inline
      tableStrategy:
        standard:
          shardingColumn: order_id
          shardingAlgorithmName: t_order_inline
```

```yaml
shardingAlgorithms:
  database_inline:
    type: INLINE
    props:
      algorithm-expression: ds_${user_id % 2}
  t_order_inline:
    type: INLINE
    props:
      algorithm-expression: t_order_${order_id % 2}
```

In the sharding configuration file, t_order is divided into four shards, namely ds_0.t_order_0, ds_0.t_order_1, ds_1.t_order_0, and ds_1.t_order_1. We can use the PREVIEW statement to view the routing results (shown in the following code block) of SELECT * FROM t_order ORDER BY order_id:

```sql
SQL
PREVIEW SELECT * FROM t_order ORDER BY order_id;
+------------------+-------------------------------------------
------+
| data_source_name | sql
                   |
+------------------+-------------------------------------------
------+
| ds_0             | SELECT * FROM t_order_0 ORDER
BY order_id |
| ds_0             | SELECT * FROM t_order_1 ORDER
BY order_id |
| ds_1             | SELECT * FROM t_order_0 ORDER
BY order_id |
| ds_1             | SELECT * FROM t_order_1 ORDER
BY order_id |
+------------------+-------------------------------------------
------+
```

The original query statement has been rewritten into the actual executable query statement, which is routed to the ds_0 and ds_1 data sources for execution. According to the grouping of the data sources, the execution statements are divided into two groups. So, in terms of the execution statements for the same data source, how many connections do we need to open to execute a query? The max-connections-size-per-query parameter is used to control the maximum number of connections that are allowed on the same data source.

When the configured value of max-connections-size-per-query is greater than or equal to the number of all SQLs that need to be executed under the data source, ShardingSphere enables its **Limited Memory** mode. This mode allows you to stream SQL queries to limit memory usage. However, when this parameter value is less than the number of all SQLs that need to be executed under the data source, ShardingSphere enables its **Limited Connection** mode and creates a unique connection for execution. Due to this, the query result set is loaded into memory to avoid consuming too many database connections. The following diagram provides an overview in the form of an *equation* for this reasoning:

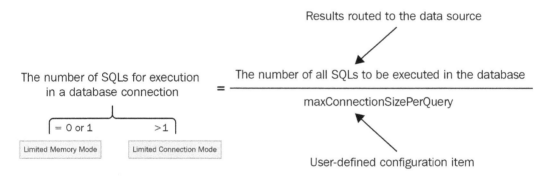

Figure 8.2 – SQL tuning

In most **Online Transactional Processing (OLTP)** scenarios, shard keys are used to ensure routing to the data node, so the value of max-connections-size-per-query is set to 1 to strictly control database connections and ensure that database resources can be used by more applications. However, in terms of **Online Analytical Processing (OLAP)** scenarios, users can configure a relatively high value of the max-connections-size-per-query parameter to improve system throughput.

Summary

After reading this chapter, you should be able to not only customize strategies for ShardingSphere's most notable features but also tune the proxy's performance. Feel free to apply the knowledge you have gathered so far, coupled with the examples offered in this chapter, to customize ShardingSphere. The examples we have provided you with are generalizable, which means that you can come back to them at any time.

If you'd like to test yourself even further, once you have completed the remaining chapters of this book, you can join the ShardingSphere developer community on GitHub, where you'll find new strategies constantly being shared by other users like you.

9
Baseline and Performance Test System Introduction

In this internet era fueled by fast-growing app-driven businesses, access to high concurrency and high throughput has become a trend. Naturally, our common goal is to provide a functioning system that can normally respond in such scenarios. So, we must master system performance skills to better deal with performance stress. In this chapter, we'd like to share our experience of performance testing with you.

This chapter will equip you with the necessary knowledge to understand and fully take advantage of ShardingSphere's baseline and performance testing. Together, we will cover the following topics:

- Baseline and benchmarking tools and application scenarios
- Performance testing – from preparing your test to report analysis

By the end of this chapter, you'll have developed an understanding of how to test your databases, what tools are at your disposal, and how ShardingSphere can simplify your testing.

Technical requirements

No hands-on experience in a specific language is required, but it would be beneficial if you have some experience in Java as ShardingSphere is coded in Java.

To run the practical examples in this chapter, you will need the following tools:

- **A 2 cores 4 GB machine with Unix or Windows OS**: ShardingSphere can be launched on most OSs.

- **JRE or JDK 8+**: This is the basic environment for all Java applications.

- **Text editor** (**not mandatory**): You can use Vim or VS Code to modify the YAML configuration files.

- **A MySQL/PG client**: You can use the default CLI or other SQL clients such as Navicat or DataGrip to execute SQL queries.

- **7-Zip or tar command**: You can use these tools for Linux or macOS to decompress the proxy artifact.

- Performance testing tools as mentioned throughout this chapter.

> **You can find the complete code file here:**
> ```
> https://github.com/PacktPublishing/A-Definitive-
> Guide-to-Apache-ShardingSphere
> ```

Baseline

The first of the two main sections of this chapter will get you started with baseline testing. You might already be familiar with the concept of software baseline testing. If not, it is very simple. **Baseline testing** refers to setting the bases of what constitutes the *average* performance of a system – it consists of setting the basis for eventual performance tests to be carried out. This will give you something to compare the performance test results with, by creating a reference to come back to in the future.

We should clarify from the beginning that baseline testing is not related to the functionality of your system or, in this case, the ShardingSphere ecosystem.

The following section will guide you through the tools that are available to perform your baseline testing, their features and application scenarios, and finally, provide you with some examples. Let's start with the available benchmarking tools.

Benchmarking tools

In the software world, you will find that there are a large number of benchmarking tools available. Some of them claim to be the best alternative, with some being commercial, while others are open source.

To simplify your benchmarking tool selection process, we have shortlisted the best alternatives and present them to you in the following sections. We have chosen these tools based on our experience.

Sysbench

Sysbench is a cross-platform multithreaded testing tool that is open source, modular, and based on **Lua**, which is a lightweight programming language that is suitable for embedded use in applications. As a testing tool, Lua evaluates performance CPU usage, I/O, memory usage, and database performance parameters. It is often used for database benchmarking. Currently, it supports MySQL, Oracle, openGauss, and PostgreSQL.

Sysbench features at your disposal

You will have probably guessed that a benchmarking tool's function is to, well, benchmark performance. But specifically, what does that mean? And what features can you expect to get access to with Sysbench? The following list of points will give you an overview of the features that you'll be able to use:

- It provides a number of speed and latency statistics (including latency percentages and histograms).

- Even with thousands of concurrent threads, the overhead is low. Sysbench can generate and track hundreds of millions of events per second.

- New benchmarks can be easily created with the implementation of predefined hooks that users supply in Lua scripts.

- Sysbench also works as a Lua interpreter. To use it, users only need to replace `#!/usr/bin/lua` with `#!/usr/bin/sysbench` in the scripts.

Application scenarios

In the previous section, we introduced you to the features of Sysbench. In this section, we will use an example to help you better assimilate the following application scenarios of said features:

- Database queries

- Read/write operations

- Insert and update operations
- Delete operations

The example we have prepared for you is useful as it is generalizable and includes the four previously cited application scenarios.

Let's go ahead and examine how you would test your baseline for these application scenarios.

For MySQL transaction-oriented **online transaction processing** (**OLTP**) testing, there are three phases: preparation, running, and results, as follows:

1. **Preparation**: During this stage, a table with a specified number of rows is created in the database. The default table name is `test` (whose default storage engine in Sysbench is `innodb`). For example, the following code is written to create a table with 1 million data rows:

```Shell
sysbench oltp_read_only --mysql-host=${IP} --mysql-
port=${MySQL_Port} --mysql-user=${MySQL_User} --mysql-
password=${MySQL_Passwd} --mysql-db=test --tables=1
--table-size=1000000 --report-interval=10 --time=3600
--threads=10 --max-requests=0 --percentile=99 --mysql-
ignore-errors="all" --rand-type=uniform --range_
selects=off --auto_inc=off prepare
```

2. **Test running**: Run an OLTP test on the `sbtest` table that we just created:

```Shell
sysbench oltp_read_only --mysql-host= ${IP} --mysql-
port= ${MySQL_Port} --mysql-user= ${MySQL_User} --mysql-
password= ${MySQL_Passwd} --mysql-db=test --tables=1
--table-size=1000000 --report-interval=16 --time=3600
--threads=10 --max-requests=0 --percentile=99 --mysql-
ignore-errors="all" --rand-type=uniform --range_
selects=off --auto_inc=off run
```

3. **Output results**: Once you run the test, you will receive a result similar to the following output:

```YAML
SQL statistics:
      # Total transactions(transactions per second)
    transactions:                            1890956
```

```
(15755.04 per sec.)
    # Total operations(operations per second)
    queries:                             11345736
(94530.25 per sec.)
Latency (ms):
    # Minimum latency
        min:                                 1.77
    # Average latency
        avg:                                 4.06
    # Maximum latency
        max:                               303.13
    # Over 99% average time
        99th percentile:                     5.88
        sum:                         7678839.72
```

This concludes our introduction of the first tool at your disposal for baseline testing, Sysbench. In the next section, we will introduce you to the second tool that we would like to recommend for baseline testing.

BenchmarkSQL

BenchmarkSQL is an open source database testing tool based on Java that includes `TPC-C` test scripts and supports the performance stress testing of databases such as MySQL, Oracle, PostgreSQL, openGauss, and SQL Server.

Application scenario

BenchmarkSQL leverages JDBC to test TPC-C for OLTP. The usage of BenchmarkSQL is similar to that of Sysbench: first, generate stress testing metrics, then execute stress tests, and output the results.

In addition to textual output, BenchmarkSQL can also use its own scripts to generate reports in HTML format (note that R dependencies will be installed first).

Let's go through an application scenario that is very useful as it is generalizable. The following example shows you how to perform baseline tests in a scenario, including transaction simulations for e-commerce systems. As you can imagine, these systems require high concurrency and high availability and are part of the fast-growing app-driven businesses we mentioned at the beginning of this chapter. For this scenario, let's start from the very beginning, by modifying the source code of BenchmarkSQL in the first step.

Modify the source code of BenchmarkSQL:

1. Modify jTPCC.java and add the MySQL-related parts:

```Java
if (iDB.equals("firebird"))
        dbType = DB_FIREBIRD;
¶    else if (iDB.equals("postgres"))
        dbType = DB_POSTGRES;
    else if (iDB.equals("mysql"))
        dbType = DB_UNKNOWN;
    else
    {
        log.error("unknown database type '" + iDB + "'");
        return;
    }
```

2. Modify jTPCCConnection.java and create the "AS L" alias for the SQL subquery:

```Java
stmtStockLevelSelectLow = dbConn.prepareStatement(
        "SELECT count(*) AS low_stock FROM (" +
        "    SELECT s_w_id, s_i_id, s_quantity " +
        "      FROM bmsql_stock " +
        "     WHERE s_w_id = ? AND s_quantity < ? AND s_i_id IN (" +
        "            SELECT ol_i_id " +
        "              FROM bmsql_district " +
        "              JOIN bmsql_order_line ON ol_w_id = d_w_id " +
        "               AND ol_d_id = d_id " +
        "               AND ol_o_id >= d_next_o_id - 20 " +
        "               AND ol_o_id < d_next_o_id " +
        "             WHERE d_w_id = ? AND d_id = ? " +
        "           ) " +
```

```
   "     )AS L");
break;
```

3. Create the MySQL-oriented `tpcc.mysql` script:

```Shell
db=mysql
driver=com.mysql.jdbc.Driver
conn=jdbc:mysql://127.0.0.1:3306/benchmarksql
user=benchmarksql
password=
warehouses=1
loadWorkers=4
terminals=1
runTxnsPerTerminal=10
runMins=0
limitTxnsPerMin=300
terminalWarehouseFixed=true
newOrderWeight=45
paymentWeight=43
orderStatusWeight=4
deliveryWeight=4
stockLevelWeight=4
resultDirectory=my_result_%tY-%tm-%td_%tH%tM%tS
```

4. Modify the `funcs.sh` file and add the MySQL database type:

```Shell
function setCP()
{
        ...omitted
    mysql)
        cp="../lib/mysql/*:../lib/*"
        ;;
    esac
    myCP=".:${cp}:../dist/*"
    export myCP
}
```

```
...Omitted
case "$(getProp db)" in
    firebird|oracle|postgres|mysql)
    ;;
    "") echo "ERROR: missing db= config option in
${PROPS}" >&2
    exit 1
    ;;
    *)   echo "ERROR: unsupported database type
'db=$(getProp db)' in ${PROPS}" >&2
    exit 1
    ;;
esac
```

Add the **MySQL Connector** driver; create a `mysql` folder under the `lib` directory and move the driver into this directory.

5. Modify `runDatabaseBuild.sh` and remove `extraHistID`:

```Makefile
Makefile
AFTER_LOAD="indexCreates foreignKeys buildFinish"
```

6. Run the test:

```CSS
CSS
./runDatabaseBuild.sh tpcc.mysql
./runBenchmark.sh tpcc.mysql
```

7. Output the results:

```PowerShell
PowerShell
18:58:17,071 [Thread-1] INFO jTPCC : Term-00,
# tPMC (Transactions Per Minute): The tPMC metric is the
number of new order transactions executed per minute.
TPC-C is also measured in transactions minute (tmpC).
18:58:17,071 [Thread-1] INFO jTPCC : Term-00, Measured
tpmC (NewOrders) = 136.71
# tpmTOTAL: Total transaction per minute
18:58:17,071 [Thread-1] INFO jTPCC : Term-00, Measured
tpmTOTAL = 298.81
```

This concludes the necessary steps to be performed with BenchmarkSQL for baseline testing. There is still another tool that is available. We will introduce it in the following section.

A good-to-know alternative benchmarking tool

The benchmarking tool **Java Microbenchmark Harness (JMH)** is developed by OpenJDK for code performance tuning. It is a nanosecond-precise system that is suitable for Java and other JVM-based languages.

Application scenarios

JMH is an alternative benchmarking tool, but not in the sense that it is any less good than Sysbench, for example. In the upcoming steps, you will understand in which scenarios JMH might be better suited to work and learn through an example how to get the best out of it. So, what are the advantages of JMH? The following three positives make it a good alternative:

- It's an ideal tool for when you want to know the precise execution time of a method and the correlation between the execution time and input.

- It's also a good choice when you need to compare throughputs of different implementations of the interface under a set of given conditions to find the optimal implementation.

- It can be used to see percentages of requests completed within given times.

Now, let's jump to the example that will help you get a feel of how JMH works. With the following example, you'll learn how to use JMH to perform a stress test on JDBC:

1. Add the JMH dependency to the pom.xml file:

    ```html
    HTML
        <properties>
            <jmh.version>1.21</jmh.version>
        </properties>
        <dependencies>
            <!-- JMH-->
            <dependency>
                <groupId>org.openjdk.jmh</groupId>
                <artifactId>jmh-core</artifactId>
                <version>${jmh.version}</version>
            </dependency>
    ```

```
        <dependency>
            <groupId>org.openjdk.jmh</groupId>
            <artifactId>jmh-generator-annprocess</
artifactId>
            <version>${jmh.version}</version>
            <scope>provided</scope>
        </dependency>
    </dependencies>
```

2. Write the test method:

```Java
@Setup(Level.Trial)
public void setup() throws Exception {
    connection = getConnection();
    for (int i = 0; i < preparedStatements.length; i++) {
        preparedStatements[i] = connection.
prepareStatement(String.format("select c from sbtest%d
where id = ?", i + 1));
    }
}


@Benchmark
public void oltpPointSelect() throws Exception {
...omitted
}


@TearDown(Level.Trial)
```

3. Compile the code and package it:

```Apache
mvn clean package
java -Dconf=config.properties -jar benchmarks.jar
UnpooledFullPointSelectBenchmark -f 3 -i 5 -r 5 -t 1 -w 3
-wf 1 -wi 1 > log.txt
```

4. Output the results:

```YAML
```

```
Result: 437959.831 ±(99.9%) 6719.199 ops/s [Average]
# Result Mean value and margin of error
  Statistics: (min, avg, max) = (426614.397, 437959.831,
448490.747), stdev = 6285.143
# min (Minimum), avg (average), max (maximum), stdev
(standard deviation)
  Confidence interval (99.9%): [431240.632, 444679.030]
# Run complete. Total time: 00:02:00
# Samples Total executions, Score average value; Error
means the margin of error
Benchmark
   Mode  Samples     Score      Error   Units
i.w.j.j.UnpooledFullPointSelectBenchmark.FullPointSelect
   thrpt      15  437959.831 ± 6719.199  ops/s
```

Before moving on to performance testing, a clarification is due when it comes to the type of database on which to execute baseline testing.

Databases

Now, let's move on to the *what's being tested* part of the equation. Benchmarking tools test system performance, and systems are built with databases. If you want to know what the differences are between the most popular offerings or just refresh your memory, the following sections will help you do just that.

MySQL

MySQL is one of the most popular relational database management systems, and it is also one of the best pieces of relational database management software for web applications. The SQL language of MySQL is the most commonly used standard language for accessing databases. MySQL software adopts a dual-licensing policy, and it has a community version and a commercial version. Given its features, such as small size, fast speed, the **low total cost of ownership** (TCO), and, in particular, being open source, MySQL is generally chosen by developers of small to medium-sized websites.

PostgreSQL

PostgreSQL is a very powerful, open source relational database management system. Most SQL standards are supported, and a multitude of other features such as complex queries, foreign keys, triggers, views, transaction integrity, and multi-version concurrency control are offered.

PostgreSQL can be extended in many ways. Some possibilities include adding new data types, functions, operators, aggregate functions, indexing methods, and procedural languages.

You now have an overview of the tools required to undertake baseline testing, along with the most important components of your system. Now, it is time to move on to how you'd undertake baseline testing with Apache ShardingSphere.

ShardingSphere

ShardingSphere includes multiple components and the corresponding testing tools that are required to run tests in a system using ShardingSphere-JDBC or ShardingSphere-Proxy:

Testing tools	ShardingSphere-JDBC	ShardingSphere-Proxy
JMH	Yes (cases need to be added)	Yes (cases need to be added)
Sysbench	No	Yes
BenchmarkSQL	Yes (needs to be modified)	Yes

Table 9.1

JMH is simply a bench testing tool without any built-in test cases. However, by adding corresponding test cases, ShardingSphere-JDBC and ShardingSphere-Proxy, which we introduced in *Chapter 5*, *Exploring ShardingSphere Adaptors*, can be tested.

Sysbench, a commonly used testing tool, includes some built-in database testing scripts. But since it uses the C language database, only ShardingSphere-Proxy can be tested.

BenchmarkSQL, the Java implementation of TPCC, focuses more on transaction performance and can test ShardingSphere-Proxy after compiling and packaging. If you want to test ShardingSphere-JDBC, first, you will need to replace the data source in its source code.

Various functions

ShardingSphere is equipped with functions such as read/write splitting, data sharding, shadow database, data encryption, and data decryption. The performance of these functions can be observed with tools such as JMH, Sysbench, and BenchmarkSQL, as mentioned earlier.

Taking ShardingSphere-Proxy plus encryption and decryption as an example, stress testing can be run through Sysbench when the following configurations are set and the proxy is launched: `config-encrypt.yaml`:

```
HTTP
schemaName: encrypt_db
```

```
dataSources:
  ds_0:
    url: jdbc:mysql://127.0.0.1:3306/encrypt_
db?serverTimezone=UTC&useSSL=false
    username: root
    password:
    connectionTimeoutMilliseconds: 30000
    idleTimeoutMilliseconds: 60000
    maxLifetimeMilliseconds: 1800000
    maxPoolSize: 50
    minPoolSize: 1

rules:
- !ENCRYPT
  encryptors:
    aes_encryptor:
      type: AES
      props:
        aes-key-value: 123456abc
    md5_encryptor:
      type: MD5
  tables:
    sbtest1:
      columns:
        pad:
          cipherColumn: pad
          encryptorName: aes_encryptor
  ...omitted     sbtest10:
      columns:
        pad:
          cipherColumn: pad
          encryptorName: aes_encryptor
  queryWithCipherColumn: true
```

Thanks to the previous example, you have learned how to perform performance testing with ShardingSphere-Proxy with the addition of the encryption and decryption feature. The next section will further enhance your capabilities by giving you the tool to be able to determine which database or middleware to choose according to their respective performances.

Performance testing

Performance testing refers to the test of a system's various performance metrics by using automated testing tools to simulate multiple normal, peak, and abnormal load conditions.

Load testing and **stress testing** are both considered performance testing and can be combined. Determining system performance under different workloads through load testing aims to help you understand the change of various system performance metrics when the load is gradually increasing. Stress testing determines the maximum service that a system can provide by testing a system's bottleneck or intolerable performance point.

Performance is critical when it comes to choosing a database or middleware product. To understand the basic performance of various types of databases, you need to run performance testing on these databases or middleware.

In the following section, you will find a complete guide on how to prepare, execute, and analyze your performance stress testing report.

Test preparation

Taking an OLTP database as an example, here are the components needed to create a preparation environment for benchmark testing:

Role	Component	Version
Stress testing tools	Sysbench	1.0.20
Database	MySQL	5.7.26
Database	PostgreSQL	14.1

Table 9.2

If you'd like to recreate our testing environment, you can easily do so. The tools we use are all very easy to find and obtain (`https://centos.pkgs.org/8/epel-x86_64/sysbench-1.0.20-5.el8.x86_64.rpm.html`, `https://dev.mysql.com/downloads/mysql/5.7.html` `https://www.postgresql.org/about/news/postgresql-141-135-129-1114-1019-and-9624-released-2349/`). Without further ado, the next section will guide you through the necessary workflow step by step.

Applying the performance testing workflow

The workflow to be performed is fairly straightforward. It just includes two steps that are provided to you in the following paragraphs.

In the following steps, we will provide you with a quick start guide to make sure there won't be any hiccups while installing Sysbench, MySQL, or PostgreSQL, Moreover, we will also give you a few additional tests if you'd like to know more about your system:

1. Use the stress testing tool to import the testing data into the database:

```Shell
sysbench oltp_read_only --mysql-host=${IP} --mysql-
port=${MySQL_Port} --mysql-user=${MySQL_User} --mysql-
password=${MySQL_Passwd} --mysql-db=${MySQL_SCHEMA}
--tables=10 --table-size=1000000 --report-interval=10
--time=3600 --threads=10 --max-requests=0 --percentile=99
--mysql-ignore-errors="all" --rand-type=uniform --range_
selects=off --auto_inc=off cleanup
sysbench oltp_read_only --mysql-host=${IP} --mysql-
port=${MySQL_Port} --mysql-user=${MySQL_User} --mysql-
password=${MySQL_Passwd} --mysql-db=${MySQL_SCHEMA}
--tables=10 --table-size=1000000 --report-interval=10
--time=3600 --threads=10 --max-requests=0 --percentile=99
--mysql-ignore-errors="all" --rand-type=uniform --range_
selects=off --auto_inc=off prepare
```

2. Execute the following stress testing command to run tests on a database:

```Shell
sysbench oltp_read_only --mysql-host=${IP} --mysql-
port=${MySQL_Port} --mysql-user=${MySQL_User} --mysql-
password=${MySQL_Passwd} --mysql-db=${MySQL_SCHEMA}
--tables=10 --table-size=1000000 --report-interval=10
--time=3600 --threads=10 --max-requests=0 --percentile=99
--mysql-ignore-errors="all" --rand-type=uniform --range_
selects=off --auto_inc=off run
```

As you can see from the previous two steps, the procedure is very simple. Once completed, you can move on to the environment deployment phase of the performance stress testing. The next section will guide you through how to successfully complete this phase.

Environment deployment

To prepare your environment, you should prepare a testing tool and the database. This preparation is outlined in the following sections.

Sysbench installation

You can very quickly complete the installation of Sysbench using the following code.

You can refer to `https://github.com/akopytov/sysbench#rhelcentos`:

```Shell
yum -y install make automake libtool pkgconfig libaio-devel
# For MySQL support, replace with mysql-devel on RHEL/CentOS 5
yum -y install mariadb-devel openssl-devel
curl -s https://packagecloud.io/install/repositories/akopytov/
sysbench/script.rpm.sh | sudo bash
sudo yum -y install sysbench
```

MySQL installation

To test MySQL, you can quickly initiate a MySQL instance via Docker:

```Shell
docker run -itd -e MYSQL_ROOT_PASSWORD=root -p3306:3306
mysql:5.7
```

PostgreSQL installation

Additionally, users can quickly initiate a PostgreSQL instance via Docker:

```Shell
docker run --name postgres -ePOSTGRES_PASSWORD=PostgreSQL@123
-p5432:5432  -d postgres
```

Stress testing

Once you have completed the environment preparation, you can perform the actual stress test. The following code will clean and prepare your data and then perform the test:

```Shell
# Clear data
sysbench oltp_read_only --mysql-host=127.0.0.1 --mysql-
port=3306 --mysql-user=root --mysql-password=root --mysql-
```

```
db=sbtest --tables=10 --table-size=1000000 --report-interval=10
--time=3600 --threads=10 --max-requests=0 --percentile=99
--mysql-ignore-errors="all" --rand-type=uniform --range_
selects=off --auto_inc=off cleanup
# Prepare data
sysbench oltp_read_only --mysql-host=127.0.0.1 --mysql-
port=3306 --mysql-user=root --mysql-password=root --mysql-
db=sbtest --tables=10 --table-size=1000000 --report-interval=10
--time=3600 --threads=10 --max-requests=0 --percentile=99
--mysql-ignore-errors="all" --rand-type=uniform --range_
selects=off --auto_inc=off prepare
# Test by using the built-in script of oltp_read_only
sysbench oltp_read_only --mysql-host=127.0.0.1 --mysql-
port=3306 --mysql-user=root --mysql-password=root --mysql-
db=sbtest --tables=10 --table-size=1000000 --report-interval=10
--time=3600 --threads=10 --max-requests=0 --percentile=99
--mysql-ignore-errors="all" --rand-type=uniform --range_
selects=off --auto_inc=off run
Additional tests
```

Based on the Sysbench stress testing tool, multiple other scenarios are provided to test performance under different scenarios, such as read-only, write-only, and read/write.

Here is the code for the read-only scenario, `oltp_point_select.lua`:

```Shell
sysbench oltp_point_select --mysql-host=127.0.0.1 --mysql-
port=3306 --mysql-user=root --mysql-password=root --mysql-
db=sbtest --tables=10 --table-size=1000000 --report-interval=10
--time=3600 --threads=10 --max-requests=0 --percentile=99
--mysql-ignore-errors="all" --rand-type=uniform --range_
selects=off --auto_inc=off run
```

Here is the code for the read/write scenario, `oltp_read_write.lua`:

```Shell
sysbench oltp_read_write --mysql-host=127.0.0.1 --mysql-
port=3306 --mysql-user=root --mysql-password=root --mysql-
db=sbtest --tables=10 --table-size=1000000 --report-interval=10
--time=3600 --threads=10 --max-requests=0 --percentile=99
--mysql-ignore-errors="all" --rand-type=uniform --range_
selects=off --auto_inc=off run
```

Here is the code for the write-only scenario, `oltp_write_only.lua`:

```Shell
sysbench oltp_write_only --mysql-host=127.0.0.1 --mysql-
port=3306 --mysql-user=root --mysql-password=root --mysql-
db=sbtest --tables=10 --table-size=1000000 --report-interval=10
--time=3600 --threads=10 --max-requests=0 --percentile=99
--mysql-ignore-errors="all" --rand-type=uniform --range_
selects=off --auto_inc=off run
```

Result report analysis

In the following code, we can view the analysis of the Sysbench result:

```Shell
    transactions:                         1294886 (21579.74 per
sec.)     # Total transactions (transactions per second)
    queries:                              1294886 (21579.74 per
sec.)     # Read total (read per second)
    Latency (ms):
        min:                                          0.36
            # Minimum delay
        avg:                                          0.74
            # Average delay
        max:                                          8.90
            # Maximum delay
        95th percentile:                              1.01
            # 95th-percentile
        sum:                                     959137.19
```

The preceding output is an example of the type of report that you can expect to be receiving at the end of your performance testing. It will allow you to determine whether your system is performing at your expected performance requirement.

Summary

Performance tests are applied to multiple scenarios, requiring various testing tools and comparison products. It is crucial to choose the appropriate tool to conduct performance tests. Sysbench and other tools are currently available for database performance benchmarking tests. Hopefully, the examples in this chapter will help you gain a basic understanding of performance testing.

You are now able to understand the difference between baseline and performance testing and select the most appropriate tool. Moreover, you are now equipped with the necessary code to perform your own tests and analyze whether your system is running according to your expectations.

In the next chapter, we will help you take your testing skills to the next level, by introducing you to scenario testing.

10
Testing Frequently Encountered Application Scenarios

This chapter will guide you through all the possible application scenarios you may encounter, thus proving to be not only a useful introductory guide to you now as a first-time user but also a long-term reference. You will start with distributed databases and work your way through the remaining major scenarios for which you might need Apache ShardingSphere. In this chapter, we will cover the following topics:

- Testing distributed database scenarios
- Scenario-based testing for database security
- Synthetic monitoring
- Database gateways

By the end of the chapter, you'll be equipped to handle testing every single one of the aforementioned scenarios, from preparation work to deployment and report analysis. The following section will teach you how to distribute database traffic to multiple databases to increase services' stability and reliability, how to distribute database traffic between production and stress testing databases, and more.

Technical requirements

No hands-on experience in a specific language is required but it would be beneficial to have some experience in Java since ShardingSphere is coded in Java.

To run the practical examples in this chapter, you will need the following tools:

- **A 2 cores 4 GB machine with Unix or Windows OS**: ShardingSphere can be launched on most OSs.

- **JRE or JDK 8+**: This is the basic environment for all Java applications.

- **Text editor (not mandatory)**: You can use Vim or VS Code to modify the YAML configuration files.

- **A MySQL/PG client**: You can use the default CLI or other SQL clients such as Navicat or DataGrip to execute SQL queries.

- **7-Zip or tar command**: You can use these tools for Linux or macOS to decompress the proxy artifact.

- **Sysbench**: For performance testing

> You can find the complete code file here:
> `https://github.com/PacktPublishing/A-Definitive-Guide-to-Apache-ShardingSphere`

Testing distributed database scenarios

In the context of fast-growing digital industries, enterprise data storage and transaction volume have grown rapidly as well. Traditional standalone databases find it increasingly difficult to support the surge in online access requests. To resolve this bottleneck, distributed databases are now becoming increasingly accepted by more and more enterprises as a worthy solution.

In this section, we will show how you can utilize ShardingSphere to distribute traditional standalone database traffic to multiple databases and provide more stable and reliable services.

The next section will help you get things started in testing your distributed scenario by giving you the tasks you should accomplish to prepare for the test.

Preparing to test your distributed system

To test a distributed scenario, you're required to install the following components:

Role	Component	Version
Stress testing tool	Sysbench	1.0.20
Database	MySQL	5.7.26
Middleware	ShardingSphere-Proxy	5.0.0

Table 10.1

Let's first look at the ShardingSphere-Proxy adaptor. The following sections will guide you through distributed system testing, complete with procedures, from deployment to report analysis for various features such as data sharding. Here is a breakdown of the steps that you will find in this section:

- *Deployment and configuration*

- *How to run your test*

- *Analyzing your test reports*

Let's have a look at what you need to do to get started with the deployment step, followed by configuration.

Deployment and configuration

First, download the binary package for **ShardingSphere-Proxy** from the ShardingSphere website: `https://shardingsphere.apache.org/index.html`.

Next, to get your configuration underway, start with the following binary package configuration. You will need to configure your data source, the proxy, and finally Sysbench. The following code will configure your data source:

1. First, input what you'd like to configure in the following way:

   ```
   config-sharding.yaml
   ```

2. Once you have specified what you'd like to configure, which, in this case, is the data source, you can utilize the following code to go through with the configuration:

```yaml
YAML
schemaName: sbtest
dataSources:
  ds_0:
    url: jdbc:mysql://ip1:3306/sbtest
    username: root
    password: root
    maxPoolSize: 256
    minPoolSize: 1
  ds_1:
    url: jdbc:mysql://ip2:3306/sbtest
    username: root
    password: root
    maxPoolSize: 256
    minPoolSize: 1

rules:
- !SHARDING
  tables:
    sbtest1:
      actualDataNodes: ds_${0..1}.sbtest1_${0..9}
      tableStrategy:
        standard:
          shardingColumn: id
          shardingAlgorithmName: table_inline_1

  defaultDatabaseStrategy:
    standard:
      shardingColumn: id
      shardingAlgorithmName: database_inline

  shardingAlgorithms:
    database_inline:
      type: INLINE
```

```
        props:
            algorithm-expression: ds_${id % 2}
        table_inline_1:
          type: INLINE
          props:
            algorithm-expression: sbtest1_${id % 10}
```

3. Now that you have configured your data source, proceed to configure your proxy. You can do so by using the following code. First, specify where to apply the configuration, as follows:

```
server.yaml
```

4. Once you have specified where you'd like to apply your configuration, use the following code to proceed with the configuration:

```
YAML
rules:
  - !AUTHORITY
    users:
      - root@%:root
      - sharding@:sharding
    provider:
      type: ALL_PRIVILEGES_PERMITTED
props: []
```

If you have followed the preceding steps, everything should have gone smoothly, which means you can now configure the third component for testing – Sysbench.

For more details, please read the Read Me file at https://github.com/akopytov/sysbench#rhelcentos.

5. To prepare Sysbench for testing, use the following code:

```
Shell
yum -y install make automake libtool pkgconfig libaio-devel
# For MySQL support, replace with mysql-devel on RHEL/CentOS 5
yum -y install mariadb-devel openssl-devel
```

```
curl -s https://packagecloud.io/install/repositories/
akopytov/sysbench/script.rpm.sh | sudo bash
sudo yum -y install sysbench
```

Once done with Sysbench, you will have completed the necessary testing preparation. In our case, we will use MySQL as an example.

To test MySQL, you can start a MySQL instance running in Docker, as follows:

```Shell
docker run -itd -e MYSQL_ROOT_PASSWORD=root -p3306:3306
mysql:5.7
```

After successfully starting the MySQL instance, you will be ready to run your test. The next section provides you with a detailed step-by-step approach.

How to run your testing on a distributed system

We will now use Sysbench to compare the differences when testing beyond the limits of normal traditional standalone database operations and ShardingSphere-Proxy.

In this test scenario, 256 concurrent threads of read and write operations are performed under a single table, which has a data volume of 40 million. First, we prepare the amount of data in the database in advance, then proceed with the stress testing itself, followed by a data cleaning step. We use MySQL as our example for Sysbench. The following steps will be followed by a section guiding you through how to analyze your testing report:

1. You should first prepare the data to be tested. To do so, you can use the following code:

    ```Shell
    sysbench oltp_read_write --mysql-host=DB1 --mysql-
    port=3306 --mysql-user=root --mysql-password='123456'
    --mysql-db=sbtest --tables=1 --table-size=40000000
    --report-interval=1   --time=256 --threads=256
    --max-requests=5 --percentile=99   --mysql-ignore-
    errors="all" --range_selects=off --rand-type=uniform
    --auto_inc=off prepare
    ```

2. Once done, you can directly move on to stress testing. The following script is the one you need for stress testing:

    ```Shell
    sysbench oltp_read_write --mysql-host=DB1 --mysql-
    ```

```
port=3306 --mysql-user=root --mysql-password='123456'
--mysql-db=sbtest --tables=1 --table-size=40000000
--report-interval=1  --time=256 --threads=256
--max-requests=5 --percentile=99  --mysql-ignore-
errors="all" --range_selects=off --rand-type=uniform
--auto_inc=off run
```

3. Once you have completed stress testing, you should clean your data. Here, we provide you with the script to do just that:

    Shell
    ```
    sysbench oltp_read_write --mysql-host=DB1 --mysql-
    port=3306 --mysql-user=root --mysql-password='123456'
    --mysql-db=sbtest --tables=1 --table-size=40000000
    --report-interval=1  --time=256 --threads=256
    --max-requests=5 --percentile=99  --mysql-ignore-
    errors="all" --range_selects=off --rand-type=uniform
    --auto_inc=off cleanup
    ```

Running your test on ShardingSphere-Proxy

When it comes to stress testing with ShardingSphere-Proxy, you're probably familiar with the procedure by now. The steps to follow are the same as the ones we introduced in the previous section, as you're expected to prepare the data, perform the test, clean the data, and then analyze the report.

In the following steps, we provide you with the necessary code to complete the testing procedure:

1. As for MySQL and Sysbench, you should first prepare your data. Use the following script to do so:

    Shell
    ```
    sysbench oltp_read_write --mysql-host=${Proxy_
    Host} --mysql-port=3307 --mysql-user=root --mysql-
    password='root' --mysql-db=sbtest --tables=1 --table-
    size=40000000 --report-interval=1  --time=300
    --threads=256 --max-requests=5 --percentile=99  --mysql-
    ignore-errors="all" --range_selects=off --rand-
    type=uniform --auto_inc=off prepare
    ```

2. Once done, proceed to the stress-testing step, with the following code:

```Shell
sysbench oltp_read_write --mysql-host=${Proxy_
Host} --mysql-port=3307 --mysql-user=root --mysql-
password='root' --mysql-db=sbtest --tables=1 --table-
size=40000000 --report-interval=1  --time=300
--threads=256 --max-requests=5 --percentile=99  --mysql-
ignore-errors="all" --range_selects=off --rand-
type=uniform --auto_inc=off run
```

3. The results you obtain should be cleaned by using the following code:

```Shell
sysbench oltp_read_write --mysql-host=${Proxy_Host}
--mysql-port=3307 --mysql-user=root --mysql-
password='root' --mysql-db=sbtest --tables=1 --table-
size=40000000 --report-interval=1  --time=300
--threads=256 --max-requests=5 --percentile=99  --mysql-
ignore-errors="all" --range_selects=off --rand-
type=uniform --auto_inc=off cleanup
```

Now that you have finished cleaning your data, you will want to understand your report results. The next section will provide you with the necessary examples.

Report analysis

In the following code snippets, the results of the analyses are presented for both MySQL and ShardingSphere-Proxy. This allows you to quickly differentiate the two at a glance.

The following output is a good example of the test results for the standalone MySQL database read/write. Take a few minutes to compare these results with the following ones for ShardingSphere-Proxy:

```Shell
sysbench 1.0.20 (using bundled LuaJIT 2.1.0-beta2)
Running the test with following options:
Number of threads: 256
Report intermediate results every 10 second(s)
Initializing random number generator from current time
Initializing worker threads...
Threads started!
```

```
[ 10s ] thds: 256 tps: 1720.38 qps: 21532.38 (r/w/o:
21532.38/0.00/0.00) lat (ms,95%): 1.04err/s: 0.00reconn/s: 0.00
...
[ 300s ] thds: 256 tps: 1920.56 qps: 21597.56 (r/w/o:
21597.56/0.00/0.00) lat (ms,95%): 1.01err/s: 0.00reconn/s: 0.00
SQL statistics:
    queries performed:
        read:                          21082460
        write:                         8432984
        other:                         4216492
        total:                         33731936
    transactions:                      2108246 (1820.86 per
sec.)
    queries:                           84306522 (281021.74
per sec.)
    ignored errors:                    0         (0.00 per sec.)
    reconnects:                        0         (0.00 per sec.)
```

The test results for the ShardingSphere-Proxy distributed read/write environment are as follows:

```
Shell
sysbench 1.0.20 (using bundled LuaJIT 2.1.0-beta2)
Running the test with following options:
Number of threads: 256
Report intermediate results every 10 second(s)
Initializing random number generator from current time
Initializing worker threads...
Threads started!
[ 10s ] thds: 256 tps: 21712.38 qps: 41532.38 (r/w/o:
21532.38/0.00/0.00) lat (ms,95%): 1.04err/s: 0.00reconn/s: 0.00
...
[ 300s ] thds: 256 tps: 22814.56 qps: 41597.56 (r/w/o:
21597.56/0.00/0.00) lat (ms,95%): 1.01 err/s: 0.00reconn/s:
0.00
SQL statistics:
    queries performed:
        read:                          41023410
```

```
        write:                          8432984
        other:                          4216492
        total:                          53731936
    transactions:                       6543300 (21811.86 per
sec.)
    queries:                            144306300 (481021.74
per sec.)
    ignored errors:                     0        (0.00 per sec.)
    reconnects:                         0        (0.00 per sec.)
```

As you can see from the previous result outputs that we prepared for you, the difference between MySQL and ShardingSphere-Proxy is a stark one. To provide you with more information and guidance, the next section will present you with another example, which includes data sharding.

Analyzing a ShardingSphere-Proxy data display – the sharding feature

Let's now look at an example including a popular feature, data sharding. The proxy is used to insert data and, based on the sharding rules, data is routed to the actual underlying data sources:

1. For our data sharding feature example, let's first start by inserting the data through ShardingSphere-Proxy. We are providing you with the output results of this example:

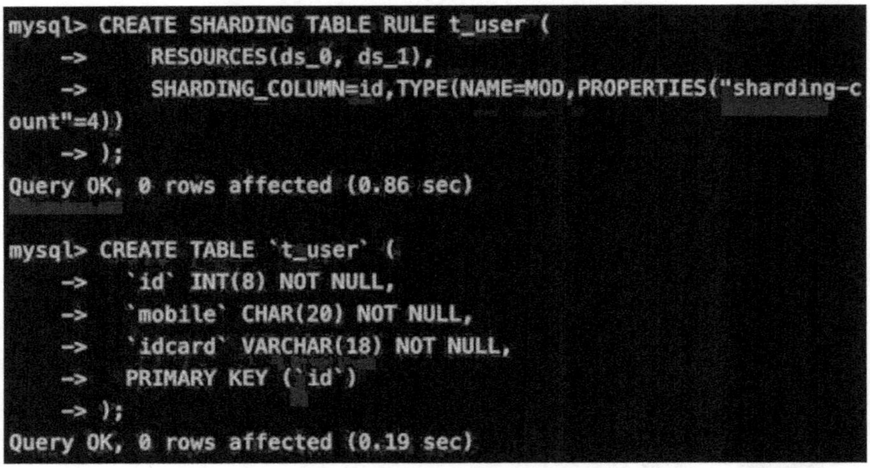

```
mysql> CREATE SHARDING TABLE RULE t_user (
    ->      RESOURCES(ds_0, ds_1),
    ->      SHARDING_COLUMN=id,TYPE(NAME=MOD,PROPERTIES("sharding-c
ount"=4))
    -> );
Query OK, 0 rows affected (0.86 sec)

mysql> CREATE TABLE `t_user` (
    ->    `id` INT(8) NOT NULL,
    ->    `mobile` CHAR(20) NOT NULL,
    ->    `idcard` VARCHAR(18) NOT NULL,
    ->    PRIMARY KEY (`id`)
    -> );
Query OK, 0 rows affected (0.19 sec)
```

Figure 10.1 – Inserting data through ShardingSphere-Proxy

2. The following presents the results of what you would be sharding if you followed the example of the previous step:

```
mysql> SELECT * FROM t_user ORDER BY id;
+----+-------------+--------------------+
| id | mobile      | idcard             |
+----+-------------+--------------------+
|  1 | 18236483857 | 220605194709308170 |
|  2 | 15686689114 | 360222198806088804 |
|  3 | 14523360225 | 411601198601098107 |
|  4 | 18143924353 | 540228199804231247 |
|  5 | 15523349333 | 360924195311103360 |
|  6 | 13261527931 | 513229195302236086 |
|  7 | 13921892133 | 500108194806107214 |
|  8 | 15993370854 | 451322194405305441 |
|  9 | 18044280924 | 411329199808285772 |
| 10 | 13983621809 | 430204195612042092 |
| 11 | 18044270924 | 411329199708285772 |
| 12 | 13983631809 | 430204195611042092 |
+----+-------------+--------------------+
12 rows in set (0.04 sec)
```

Figure 10.2 – Result of inserting data through ShardingSphere-Proxy

3. Once you have sharded your logical database, you will receive an output similar to the following figures presenting the data that has just been sharded across multiple databases. Each figure represents the output in an actual physical database:

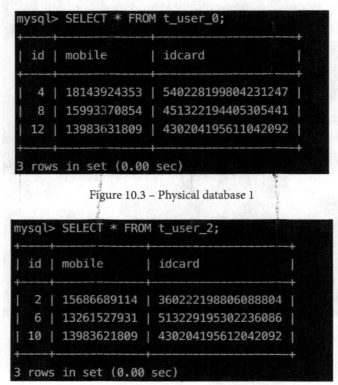

Figure 10.3 – Physical database 1

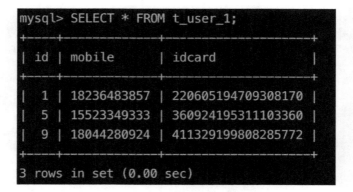

Figure 10.4 – Physical database 2

```
mysql> SELECT * FROM t_user_1;
+----+------------+--------------------+
| id | mobile     | idcard             |
+----+------------+--------------------+
|  1 | 18236483857 | 220605194709308170 |
|  5 | 15523349333 | 360924195311103360 |
|  9 | 18044280924 | 411329199808285772 |
+----+------------+--------------------+
3 rows in set (0.00 sec)
```

Figure 10.5 – Physical database 3

```
mysql> SELECT * FROM t_user_3;
+-----+-------------+--------------------+
| id  | mobile      | idcard             |
+-----+-------------+--------------------+
|  3  | 14523360225 | 411601198601098107 |
|  7  | 13921892133 | 500108194806107214 |
| 11  | 18044270924 | 411329199708285772 |
+-----+-------------+--------------------+
3 rows in set (0.00 sec)
```

Figure 10.6 – Physical database 4

Let's now jump into what the results presented to you in the previous figures mean. The following section will elaborate on these results, teaching you how to read an analysis report.

Report analysis

Based on the previous comparison, we can conclude the following from our findings:

- When there are 256 threads and 40 million data rows in a single table of the standalone MySQL database, the **transactions per second** (**TPS**) metric in the read/write stress test is only 1,820.

- However, the sharding function of ShardingSphere-Proxy can distribute 40 million data rows into the underlying five databases, and thus the TPS result is greatly increased to 21,811.

The excellent TPS numbers stand as a testament to ShardingSphere's capability to solve the bottleneck affecting traditional standalone databases, and to provide better services.

The next section will introduce you to an essential scenario that you are bound to be working with eventually at some point – database security.

Scenario-based testing for database security

Data security has been put under increased scrutiny in recent years, stemming from concerns related to the development of an increasingly internet-connected society. In the era of big data, privacy is a serious issue, and we must secure the transmission of data. This is a fundamental section for you, as when working with data, whether you are a **database administrator** (**DBA**) or an operations software engineer, you will surely dedicate time and effort to data security.

In this section, we will show how ShardingSphere can encrypt data to provide stable and reliable services.

The procedure is the same as the one that we outlined for you in the previous section, starting with a preparation phase followed by deployment, configuration, running of the test, and finally, analyzing the report.

Preparing to test your database security

To perform the test, please prepare the following components:

Role	Component	Version
Database	MySQL	5.7.26
Middleware	ShardingSphere-Proxy	5.0.0

Table 10.2

Deployment and configuration

Next, download the binary package for ShardingSphere-Proxy from the ShardingSphere website at `https://shardingsphere.apache.org/index.html`.

For the configuration of your database security testing, refer to the following code presented in this section. We have prepared an example that easily fits all cases. The first step of this example is to configure the encryption feature:

```
config-encrypt.yaml
schemaName: encrypt_db
dataSources:
  ds_0:
    url: jdbc:mysql://127.0.0.1:3306/encrypt_
db?serverTimezone=UTC&useSSL=false
    username: test
    password: test
    connectionTimeoutMilliseconds: 30000
    idleTimeoutMilliseconds: 60000
    maxLifetimeMilliseconds: 1800000
    maxPoolSize: 3000
    minPoolSize: 1
rules:
```

```
-  !ENCRYPT
   encryptors:
     aes_encryptor:
       type: AES
       props:
         aes-key-value: 123456abc
     md5_encryptor:
       type: MD5
   tables:
     sbtest1:
       columns:
         pad:
           cipherColumn: pad
           encryptorName: aes_encryptor
```

Once you have configured the encryption feature, you can test it in Sysbench as shown previously in this chapter, and you will get a similar output result to the following:

```
Shell
sysbench 1.0.20 (using bundled LuaJIT 2.1.0-beta2)
Running the test with following options:
Number of threads: 256
Report intermediate results every 10 second(s)
Initializing random number generator from current time
Initializing worker threads...
Threads started!
[ 10s ] thds: 256 tps: 21712.38 qps: 41532.38 (r/w/o:
21532.38/0.00/0.00) lat (ms,95%): 1.04err/s: 0.00reconn/s: 0.00
...
[ 300s ] thds: 256 tps: 22814.56 qps: 41597.56 (r/w/o:
21597.56/0.00/0.00) lat (ms,95%): 1.01 err/s: 0.00reconn/s:
0.00
SQL statistics:
    queries performed:
        read:                          41023410
        write:                         8432984
        other:                         4216492
        total:                         53731936
```

```
    transactions:                         6543300 (21811.86 per
sec.)
    queries:                              144306300 (481021.74
per sec.)
    ignored errors:                       0       (0.00 per sec.)
    reconnects:                           0       (0.00 per sec.)
YAML
rules:
  - !AUTHORITY
    users:
      - root@%:root
      - sharding@:sharding
    provider:
      type: ALL_PRIVILEGES_PERMITTED
props: []
```

To test MySQL, you can start a MySQL instance running in Docker:

```Shell
docker run -itd -e MYSQL_ROOT_PASSWORD=test -p3306:3306
mysql:5.7
```

How to run your testing on database security

First, you should utilize ShardingSphere-Proxy to encrypt data, and then, insert the data into the MySQL database:

```SQL
CREATE TABLE t_user (id INT(8), mobile VARCHAR(50), idcard
VARCHAR(50));

INSERT INTO t_user (id, mobile, idcard)
VALUES (1, 18236483857, 220605194709308170),
       (2, 15686689114, 360222198806088804),
       (3, 14523360225, 411601198601098107),
       (4, 18143924353, 540228199804231247),
       (5, 15523349333, 360924195311103360),
       (6, 13261527931, 513229195302236086),
       (7, 13921892133, 500108194806107214),
```

```
(8,  15993370854,  451322194405305441),
(9,  18044280924,  411329199808285772),
(10, 13983621809,  430204195612042092);
```

Report analysis

The following figures present you with an example output with and without ShardingSphere-Proxy. As you can see from the idcard column on the right, the data is completely exposed:

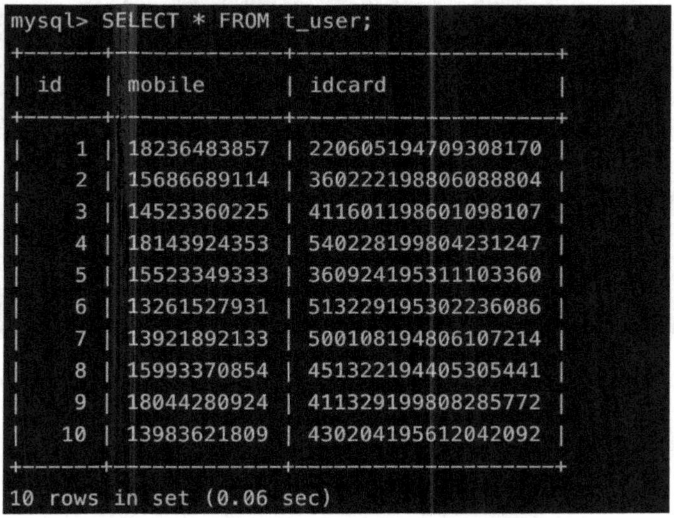

```
mysql> SELECT * FROM t_user;
+------+-------------+--------------------+
| id   | mobile      | idcard             |
+------+-------------+--------------------+
|    1 | 18236483857 | 220605194709308170 |
|    2 | 15686689114 | 360222198806088804 |
|    3 | 14523360225 | 411601198601098107 |
|    4 | 18143924353 | 540228199804231247 |
|    5 | 15523349333 | 360924195311103360 |
|    6 | 13261527931 | 513229195302236086 |
|    7 | 13921892133 | 500108194806107214 |
|    8 | 15993370854 | 451322194405305441 |
|    9 | 18044280924 | 411329199808285772 |
|   10 | 13983621809 | 430204195612042092 |
+------+-------------+--------------------+
10 rows in set (0.06 sec)
```

Figure 10.7 – Database security testing output without ShardingSphere-Proxy

Comparing the previous output with the following output, and always referring to the right column, you can see that the `idcard` data has been successfully encrypted and that the information is protected:

```
mysql> SELECT * FROM t_user;
+------+----------------------+--------------------------------------------------+
| id   | mobile_cipher        | idcard_cipher                                    |
+------+----------------------+--------------------------------------------------+
|    1 | p31Pkl9nIunYdH+AngyNUA== | pQv0JEkM94QzktJdM8UMg/uLrU71G6n6DALdPp9w6L0= |
|    2 | CV8+uYRaWOzcTQnQX3RcwA== | dCF7k4haK0aIV/d7dtwgzIb4lIFlJ913hrPim1+J278= |
|    3 | jnfu7o44KgN/PV1zhiu7jw== | 8iulp3+XTSv2XHGUUHKV0UsLuFx7yEpQVT+47EFfg94= |
|    4 | ZJDrTv/XIjdqdG1yp0t95w== | iqU6myMGfgI/XnxCtjhbMrwIauriWu8crxPS6BH2pMk= |
|    5 | FnQMYGnFJaiWmTHeNYzbFA== | KAPrCXoo1svMt5NWe0UaKYZIl1rSEVddHbBJO1jPIqw= |
|    6 | lv2ECfTCgQQksvdPp6k3Ug== | BBBPAuwU+iJluI9d9TA+H81BPnVXBaly1BE3EplN4e8= |
|    7 | z46vpnHCFTkIF2EtntxpHQ== | Bc39nPtyz1ji9Rc8k4f7G9CKfPew23mKFwp8guK7ybg= |
|    8 | p/IJdGcCikhpCu5gVZj4jg== | nnv/kS1i7uHXKncUOuLzE80WM0nGlcGkLokT2dltSaQ= |
|    9 | NvPcQv4w3EqD77+VAX0KCA== | +yeo5LWKNWcekFqYawCKjsctAZqe104DrI7AeZdR/Uk= |
|   10 | xOyg9E0X9lhy9mUx0QyL0A== | U7P1CMcxn6VPHYHPgTAtjHEbb6N6vhGOpdJtVjAdHlA= |
+------+----------------------+--------------------------------------------------+
10 rows in set (0.00 sec)
```

Figure 10.8 – Database security testing output including ShardingSphere-Proxy

Interpreting the results

As we can see from the previous figures, before including ShardingSphere-Proxy to insert data, the original data was stored in plaintext (as shown in *Figure 10.1*), but the actual data inserted into the MySQL database is displayed in ciphertext (as shown in *Figure 10.2*) based on the encryption rules of ShardingSphere-Proxy.

Let's now move on to the next scenario we have prepared for you – synthetic monitoring. The following sections will provide you with complete guidance on how to perform scenario testing for synthetic monitoring.

Synthetic monitoring

With the increasing dependence on digital services and infrastructure, the demand for data storage and business data processing capability put forward by enterprises is growing faster than ever.

In particular, surging visits to websites and apps coinciding with public holidays or impromptu marketing campaigns could compromise system performance or even completely crash systems.

Full-link stress testing can help simulate visits and simulate system execution at peak times in advance. This allows the estimation of system capacity and easily coping with visit peaks in real situations. By joining efforts, the three top open source communities (Apache APISIX, Apache ShardingSphere, and Apache SkyWalking) created CyborgFlow, a full-link stress testing solution for production.

In this section, we will introduce you to how ShardingSphere can distribute traffic to the production database and stress testing database, helping enterprises gain a clear picture of the upcoming function.

As for the previous two sections, you will find that the procedure to be followed is essentially the same, except for the script to be used.

Preparing to test synthetic monitoring

The following table consists of the components required to create a full-link test:

Role	Component	Version
Stress testing tool	Sysbench	1.0.20
Database	MySQL	5.7.26
Middleware	ShardingSphere-Proxy	5.0.0

Table 10.3

Deployment and configuration

The following sections provide you with guidance on how to perform synthetic monitoring testing with ShardingSphere-Proxy, Sysbench, and MySQL.

ShardingSphere-Proxy

Get a ShardingSphere-Proxy binary package from the official ShardingSphere website (https://shardingsphere.apache.org).

For the configuration of your database synthetic monitoring, refer to the following code we have prepared for you:

```
YAML
schemaName: sbtest
```

```yaml
dataSources:
  ds:
    url: jdbc:mysql://DB1:3306/shadow_db0
    username: root
    password: 123456
    maxPoolSize: 256
    minPoolSize: 1
  shadow_ds:
    url: jdbc:mysql://DB2:3306/shadow_db1
    username: root
    password: 123456
    maxPoolSize: 256
    minPoolSize: 1

rules:
- !SHADOW
  dataSources:
    shadowDataSource:
      sourceDataSourceName: ds
      shadowDataSourceName: shadow_ds
  tables:
    sbtest1:
      dataSourceNames:
        - shadowDataSource
      shadowAlgorithmNames:
        - user-id-insert-match-algorithm
  shadowAlgorithms:
    user-id-insert-match-algorithm:
      type: REGEX_MATCH
      props:
        operation: insert
        column: id
        regex: "[1]"
YAML
rules:
  - !AUTHORITY
```

```
    users:
        - root@%:root
        - sharding@:sharding
    provider:
        type: ALL_PRIVILEGES_PERMITTED

props: []
```

Sysbench

To use Sysbench, you can refer to the following code or the online reference (https://github.com/akopytov/sysbench#rhelcentos):

```Shell
yum -y install make automake libtool pkgconfig libaio-devel
# For MySQL support, replace with mysql-devel on RHEL/CentOS 5
yum -y install mariadb-devel openssl-devel
curl -s https://packagecloud.io/install/repositories/akopytov/
sysbench/script.rpm.sh | sudo bash
sudo yum -y install sysbench
```

MySQL

If you need to test with MySQL, a MySQL instance can be quickly launched through Docker:

```Shell
docker run -itd -e MYSQL_ROOT_PASSWORD=root -p3306:3306
mysql:5.7
```

How to run your testing on synthetic monitoring

First, obtain the derivative of ShardingSphere-Proxy (shadow) with the Sysbench tool and analyze the data distributions as follows:

```Shell
sysbench oltp_read_write --mysql-host=${Proxy_Host} --mysql-
port=3307 --mysql-user=root --mysql-password='root' --mysql-
db=sbtest --tables=1 --table-size=40000000 --report-
interval=1  --time=300 --threads=256 --max-requests=5
--percentile=99  --mysql-ignore-errors="all" --range_
```

```
selects=off --rand-type=uniform --auto_inc=off cleanup
sysbench oltp_read_write --mysql-host=${Proxy_Host} --mysql-port=3307 --mysql-user=root --mysql-password='root' --mysql-db=sbtest --tables=1 --table-size=40000000 --report-interval=1  --time=300 --threads=256 --max-requests=5 --percentile=99  --mysql-ignore-errors="all" --range_selects=off --rand-type=uniform --auto_inc=off prepare
sysbench oltp_read_write --mysql-host=${Proxy_Host} --mysql-port=3307 --mysql-user=root --mysql-password='root' --mysql-db=sbtest --tables=1 --table-size=40000000 --report-interval=1  --time=300 --threads=256 --max-requests=5 --percentile=99  --mysql-ignore-errors="all" --range_selects=off --rand-type=uniform --auto_inc=off run
```

Report analysis

The following figures (*Figures 10.9* and *10.10*) give you an example of the type of report output you will be receiving when using ShardingSphere-Proxy. We have included the resulting output for both the production and the shadow database.

The output for the production database is as follows:

```
+-------------+--------+------------+
| mobile      | status | type       |
+-------------+--------+------------+
| 13261527931 | 2      | product_ds |
| 14523360225 | 2      | product_ds |
| 15523349313 | 2      | product_ds |
| 15523349323 | 2      | product_ds |
| 15523349333 | 2      | product_ds |
| 15686689114 | 2      | product_ds |
| 18143924153 | 2      | product_ds |
| 18143924253 | 2      | product_ds |
| 18143924353 | 2      | product_ds |
| 18236483857 | 2      | product_ds |
+-------------+--------+------------+
10 rows in set (0.00 sec)
```

Figure 10.9 – Production database result output

The output for the shadow database is as follows:

```
+---------------+--------+-------------+
| mobile        | status | type        |
+---------------+--------+-------------+
| 15639784513   | 1      | shadow_ds   |
| 15639784703   | 1      | shadow_ds   |
| 15639784713   | 1      | shadow_ds   |
| 15716172114   | 1      | shadow_ds   |
| 15716173114   | 1      | shadow_ds   |
| 18099515602   | 1      | shadow_ds   |
| 18099515611   | 1      | shadow_ds   |
| 18099515622   | 1      | shadow_ds   |
| 18766746515   | 1      | shadow_ds   |
| 18766747515   | 1      | shadow_ds   |
+---------------+--------+-------------+
10 rows in set (0.00 sec)
```

Figure 10.10 – Shadow database result output

Now that you have gone through the necessary steps for testing your production and shadow database, let's see how you should analyze the report.

Interpreting the results

As you can see from the previous figures illustrating example results, we can state that, according to configuration rules, traffic can be routed to various underlying storage nodes through the shadow function of ShardingSphere, achieving the isolation of production results from stress testing results.

There now remains only one scenario. The database gateway is a frequently encountered application scenario, and the following section will guide you through performing the necessary testing.

Database gateway

For websites with high traffic that wish to deal with vast concurrent access through software, distributed load balancing on the websites is far from enough to achieve that goal.

Operations on traditional database architectures or heavy database connectivity requirements with a single server will easily cause crashes on the business layer or data access layer, or even worse, data loss.

We will have to think about how to reduce database connectivity. This section will be an introduction to this topic. We will use ShardingSphere's adaptors to show you how to achieve data routing to provide reliable services.

Preparation to test the database gateway

The components needed to perform database gateway scenario tests are as follows:

Items	Components	Version
Database	MySQL	5.7.26
Middleware	ShardingSphere-Proxy	5.0.0

Table 10.4

Deployment and configuration

The following sections provide you with guidance on how to perform synthetic monitoring testing with ShardingSphere-Proxy and MySQL.

ShardingSphere-Proxy

Download the ShardingSphere-Proxy binary package from ShardingSphere's official website (`https://shardingsphere.apache.org`).

In the following code blocks you will find the necessary code, from read/write splitting to routing:

```YAML
schemaName: sbtest

dataSources:
  write_ds:
    url: jdbc:mysql://127.0.0.1:3306/
sbtest?serverTimezone=UTC&useSSL=false
    username: root
    password: root
    maxPoolSize: 256
    minPoolSize: 1
  read_ds:
```

```yaml
    url: jdbc:mysql://127.0.0.2:3306/
sbtest?serverTimezone=UTC&useSSL=false
      username: root
      password: root
      maxPoolSize: 256
      minPoolSize: 1
rules:
- !READWRITE_SPLITTING
  dataSources:
    pr_ds:
      writeDataSourceName: write_ds
      readDataSourceNames:
        - read_ds
YAML
schemaName: test

dataSources:
  ds_0:
    url: jdbc:mysql://127.0.0.1:3306/
sbtest?serverTimezone=UTC&useSSL=false
      username: root
      password: root
      maxPoolSize: 256
      minPoolSize: 1
  ds_1:
    url: jdbc:mysql://127.0.0.2:3306/
sbtest?serverTimezone=UTC&useSSL=false
      username: root
      password: root
      maxPoolSize: 256
      minPoolSize: 1
rules:
- !READWRITE_SPLITTING
 dataSources:
   pr_ds0:
      writeDataSourceName: ds_0
```

```
        readDataSourceNames:
          - ds_0
      pr_ds1:
        writeDataSourceName: ds_1
        readDataSourceNames:
          - ds_1
YAML
rules:
  - !AUTHORITY
    users:
        - root@%:root
        - sharding@:sharding
    provider:
        type: NATIVE
props:
  proxy-frontend-executor-size: 200
  proxy-backend-executor-suitable: OLTP
  sql-show: true
```

To test MySQL, you can quickly initiate two MySQL instances via Docker:

```Shell
docker run -itd -e MYSQL_ROOT_PASSWORD=root -p3306:3306
mysql:5.7
```

How to run your testing on a database gateway

Insert data into two MySQL databases manually or automatically with the help of tools and try to insert different data for the convenience of differentiation.

> **Note**
> The primary and secondary nodes will not be used for demonstration purposes in this case.

Case 1 – Read/write splitting

In this case, we will first insert data using MySQL, as follows:

```SQL
USE write_ds;
INSERT INTO t_order (order_id, user_id, status)
VALUES (1, 10001, 'write'),
       (2, 10002, 'write'),
       (3, 10003, 'write'),
       (4, 10004, 'write'),
       (5, 10005, 'write'),
       (6, 10006, 'write'),
       (7, 10007, 'write'),
       (8, 10008, 'write'),
       (9, 10009, 'write'),
       (10, 10010, 'write');

USE read_ds;
INSERT INTO t_order (order_id, user_id, status)
VALUES (1, 20001, 'read'),
       (2, 20002, 'read'),
       (3, 20003, 'read'),
       (4, 20004, 'read'),
       (5, 20005, 'read'),
       (6, 20006, 'read'),
       (7, 20007, 'read'),
       (8, 20008, 'read'),
       (9, 20009, 'read'),
       (10, 20010, 'read');
```

Next, we will use ShardingSphere-Proxy. Log in to ShardingSphere-Proxy and make query requests. The query results are from the reading database:

```SQL
SELECT * FROM t_order;
```

Log in to ShardingSphere-Proxy, alter the read/write splitting rule, and change the reading database to the writing database. The query results are from the writing database. The coding for the change is as follows:

```SQL
ALTER READWRITE_SPLITTING RULE rw_rule
    (WRITE_RESOURCE=read_ds,READ_RESOURCES(write_ds),
TYPE(NAME=random));
```

Case 2 – Data routing with multi-nodes

We will follow the pattern from the previous section here as well. We will begin by using MySQL:

```SQL
ds_0:
CREATE TABLE `t_order` (
  `order_id` int(11) NOT NULL,
  `user_id` varchar(18) NOT NULL,
  `status` varchar(255) NOT NULL,
  PRIMARY KEY (`order_id`)
) ENGINE=InnoDB DEFAULT CHARSET=utf8mb4;

ds_1:
CREATE TABLE `t_user` (
  `id` int(8) NOT NULL,
  `mobile` char(20) NOT NULL,
  `idcard` varchar(18) NOT NULL,
  PRIMARY KEY (`id`)
) ENGINE=InnoDB DEFAULT CHARSET=utf8mb4;
```

We will move on to ShardingSphere-Proxy now. Log in to ShardingSphere-Proxy and operate as follows:

```SQL
insert into t_order values(1,1,1);
insert into t_user values(1,1,1);
select * from t_order;
select * from t_user;
```

Report analysis

The analysis of the two cases is presented in the following subsections.

Case 1 – Read/write splitting

Before altering the read/write splitting rule, the query results from the reading database are as follows:

```
mysql> SELECT * FROM t_order;
+----------+---------+--------+
| order_id | user_id | status |
+----------+---------+--------+
|        1 |   20001 | read   |
|        2 |   20002 | read   |
|        3 |   20003 | read   |
|        4 |   20004 | read   |
|        5 |   20005 | read   |
|        6 |   20006 | read   |
|        7 |   20007 | read   |
|        8 |   20008 | read   |
|        9 |   20009 | read   |
|       10 |   20010 | read   |
+----------+---------+--------+
10 rows in set (0.16 sec)
```

Figure 10.11 – Reading database query results

After altering the read/write splitting rule, the query results from the writing database are as follows:

```
mysql> SELECT * FROM t_order;
+----------+---------+--------+
| order_id | user_id | status |
+----------+---------+--------+
|        1 |   10001 | write  |
|        2 |   10002 | write  |
|        3 |   10003 | write  |
|        4 |   10004 | write  |
|        5 |   10005 | write  |
|        6 |   10006 | write  |
|        7 |   10007 | write  |
|        8 |   10008 | write  |
|        9 |   10009 | write  |
|       10 |   10010 | write  |
+----------+---------+--------+
10 rows in set (0.01 sec)
```

Figure 10.12 – Writing database query results

Case 2 – Data routing with multi-nodes

After routing data with multi-nodes, you should be expecting to see an output result that is similar to the following figure:

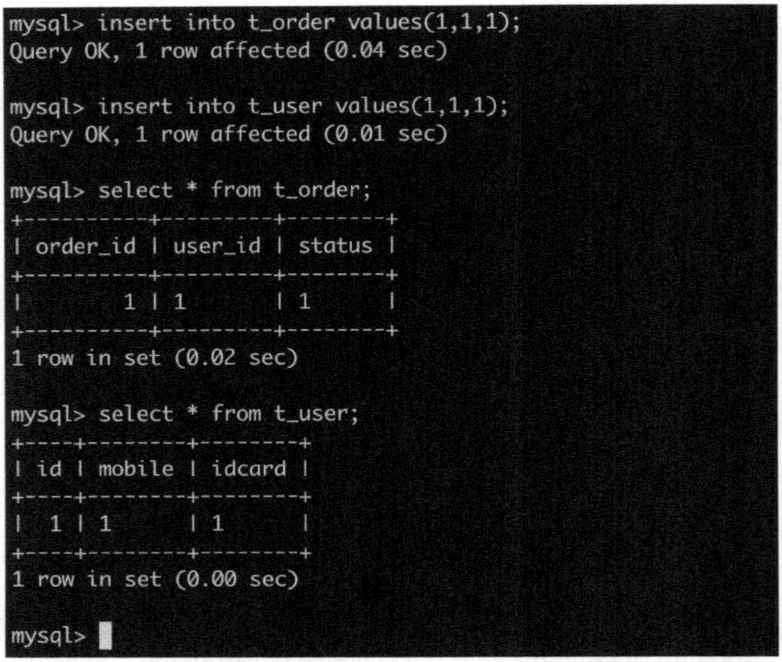

```
mysql> insert into t_order values(1,1,1);
Query OK, 1 row affected (0.04 sec)

mysql> insert into t_user values(1,1,1);
Query OK, 1 row affected (0.01 sec)

mysql> select * from t_order;
+----------+----------+--------+
| order_id | user_id  | status |
+----------+----------+--------+
|        1 | 1        | 1      |
+----------+----------+--------+
1 row in set (0.02 sec)

mysql> select * from t_user;
+----+--------+--------+
| id | mobile | idcard |
+----+--------+--------+
|  1 | 1      | 1      |
+----+--------+--------+
1 row in set (0.00 sec)

mysql>
```

Figure 10.13 – Multi-node routing data result

Check the log to see how SQL statements route:

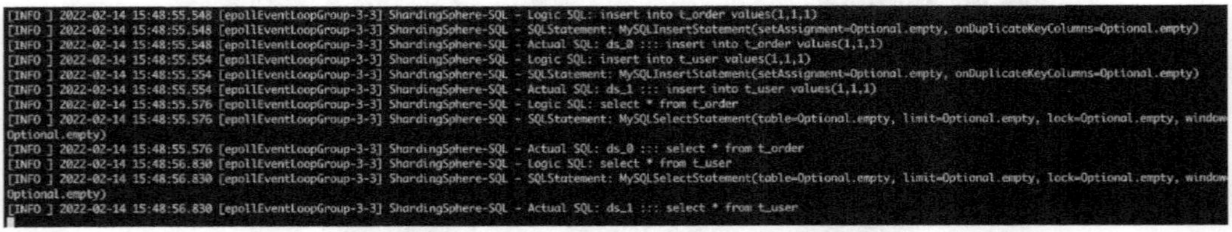

Figure 10.14 – SQL statements route log

Interpreting the results

Thanks to the previous figures, the comparison between the two cases shows us that ShardingSphere's products can realize data routing among nodes. The upper layer data can route to different storage nodes of the bottom layer according to the configured rules.

The report analysis marks the last step of your database gateway scenario testing. Coincidentally, this section also marks the end of this chapter.

Summary

By completing this chapter, you have gone through four important scenarios that you will find to be fundamental not only with Apache ShardingSphere but also when working with databases in general.

You first learned how to distribute database traffic to multiple databases to increase service stability and reliability. Then, you moved on to learn how to distribute database traffic between production and stress testing databases, test your database gateway, and finally, encrypt your data with ShardingSphere.

In the next chapter, you will come across some notable best use cases and real-world examples for using Apache ShardingSphere.

11
Exploring the Best Use Cases for ShardingSphere

You are approaching the end of this book – some might say that congratulations are in order. By now, you'll have developed a keen understanding of Apache ShardingSphere, its architecture, clients, and features, and even probably started trying it in your environment.

By all means, you could consider yourself ready to get to work by leveraging ShardingSphere. Nevertheless, we thought we would share this chapter with you. We have carefully selected some cases from real-world scenarios that a few Apache ShardingSphere users frequently deal with. These users include **small and medium enterprises (SMEs)** as well as large enterprises listed on the international stock markets.

These will help you better orient yourself among the numerous possibilities that ShardingSphere offers when paired with your databases. In this chapter, we're going to cover the following main topics:

- Recommended distributed database solution

- Recommended database security solution

- Recommended synthetic monitoring solution

- Recommended database gateway solution

Technical requirements

No hands-on experience in a specific language is required but it would be beneficial to have some experience in Java since ShardingSphere is coded in Java.

To run the practical examples in this chapter, you will need the following tools:

- **A 2 cores 4 GB machine with Unix or Windows OS**: ShardingSphere can be launched on most OSs.

- **JRE or JDK 8+**: This is the basic environment for all Java applications.

- **Text editor** (**not mandatory**): You can use Vim or VS Code to modify the YAML configuration files.

- **A MySQL/PG client**: You can use the default CLI or other SQL clients such as Navicat or DataGrip to execute SQL queries.

- **7-Zip or tar command**: You can use these tools for Linux or macOS to decompress the proxy artifact.

> **You can find the complete code file here:**
> ```
> https://github.com/PacktPublishing/A-Definitive-
> Guide-to-Apache-ShardingSphere
> ```

Recommended distributed database solution

Standalone storage and query performance bottlenecks often occur in traditional relational databases. To solve these problems, Apache ShardingSphere proposes a lightweight distributed database solution and also achieves enhanced features such as data sharding, distributed transaction, and elastic scaling, based on the storage and computing capabilities of relational databases.

With the help of Apache ShardingSphere technologies, enterprises don't need to bear the technical risks brought by replacing storage engines. On the premise that their original relational databases are stable, enterprises can still have the scalability of distributed databases.

The architecture of the distributed database solution contains five core functions, as follows:

- Sharding
- Read/write splitting
- Distributed transactions
- High availability
- Elastic scaling

You can see these core functions in *Figure 11.1*:

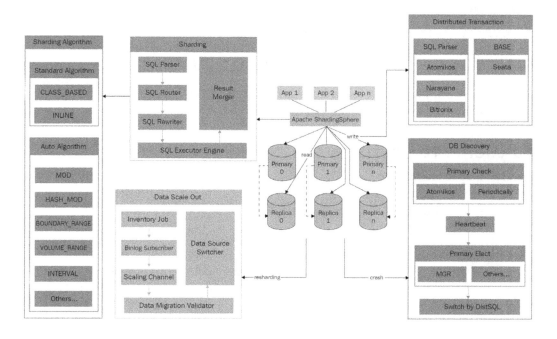

Figure 11.1 – ShardingSphere distributed database solution architecture

Data sharding allows users to define sharding rules to transparently process sharding. Apache ShardingSphere has multiple built-in sharding algorithms, such as standard sharding algorithms and auto-sharding algorithms. You can also leverage the extension point to define a custom sharding algorithm.

Based on the primary-secondary database architecture, read/write splitting can route write and read requests to the write database and read database respectively according to the user's SQL, to improve the database query performance. Concurrently, read/write splitting also eliminates the read/write lock, effectively enhancing write performance.

To manage the distributed transaction in distributed databases, Apache ShardingSphere has the built-in integration of **eXtended Architecture** (**XA**) transactions and **Basically Available, Soft state, Eventually consistent** (**BASE**) transactions. The former can meet the requirements for strong consistency, while the latter can ensure eventual consistency. By integrating the mainstream transaction solutions, ShardingSphere is enabled to meet the transaction management needs in various user scenarios.

Apache ShardingSphere's **High Availability** (**HA**) function not only utilizes the HA of the underlying database but also provides automated database discovery that can automatically sense changes in the primary and secondary database relationship, thereby correcting the connection of the computing node to the database and ensuring HA at the application layer.

Elastic scaling is the function used to scale out distributed databases. The function supports both homogeneous and heterogeneous databases, so users can directly manage scale-out with the help of DistSQL and trigger it by modifying sharding rules.

In the following sections, you will learn how to use the tools at your disposal to create your distributed databases, how to configure them, and how to use their features.

Two clients to choose from

Currently, Apache ShardingSphere's distributed database solution supports ShardingSphere-JDBC and ShardingSphere-Proxy, but it's planned to support cloud-native **ShardingSphere-Sidecar** in the future.

The access terminal of ShardingSphere-JDBC targets Java applications. Since its implementation complies with the **Java database connectivity** (**JDBC**) standards, it is compatible with mainstream **object relation mapping** (**ORM**) frameworks, such as MyBatis and Hibernate. Users can quickly integrate ShardingSphere-JDBC by installing its `.jar` packages.

Additionally, the architecture of ShardingSphere-JDBC is decentralized, so it shares resources with applications. ShardingSphere-JDBC is suitable for high-performance **online transaction processing** (**OLTP**) applications developed in Java. You can deploy ShardingSphere-JDBC in the standalone mode or the large-scale cluster mode together with the governance function to uniformly manage cluster configuration.

ShardingSphere-Proxy provides a centralized static entry, and it is suitable for heterogeneous languages and **online analytical processing (OLAP)** application scenarios. *Figure 11.2* gives you an overview of how a deployment including both ShardingSphere-JDBC and ShardingSphere-Proxy is structured:

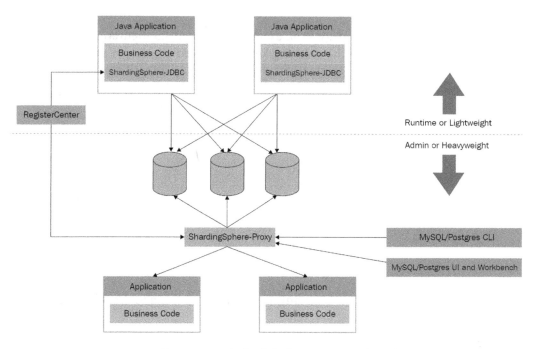

Figure 11.2 – Hybrid deployment example

ShardingSphere-JDBC and ShardingSphere-Proxy can be deployed independently, and hybrid deployment is also allowed. By adopting a unified registry center to centrally manage configurations, according to the characteristics of different access terminals, architects can build application systems suitable for all kinds of scenarios.

Your DBMS

So far, the distributed database solution of Apache ShardingSphere has supported some popular relational databases, such as MySQL, PostgreSQL, Oracle, SQLServer, and the relational databases adhering to the SQL-92 standard. Users can choose a database as the storage node of Apache ShardingSphere based on the current database infrastructure.

In the future, Apache ShardingSphere will continue to better support heterogeneous databases, including NoSQL, NewSQL, and so on. It will use the centralized database gateway as the entrance, and its internal SQL dialect converter to convert SQL into the SQL dialects of heterogeneous databases to centralize heterogeneous database management.

Sharding strategy

There are four built-in sharding strategies in Apache ShardingSphere: `Standard`, `Complex`, `Hint`, and `None`. A sharding strategy includes a **sharding key** and **sharding algorithm**, and accordingly, the user needs to specify the sharding key and sharding algorithm of the sharding strategy.

Take the `Standard` sharding strategy as an example. The usual configuration is shown as follows: `shardingColumn` is the sharding key, and `shardingAlgorithmName` is the sharding algorithm. The combination of the sharding key and sharding algorithm can implement sharding both at the database level and the table level:

```YAML
rules:
- !SHARDING
  tables:
    t_order:
      databaseStrategy:
        standard:
          shardingColumn: user_id
          shardingAlgorithmName: database_inline
```

After this, you need to configure the sharding algorithms, according to the following script:

```YAML
  shardingAlgorithms:
    database_inline:
      type: INLINE
      props:
        algorithm-expression: ds_${user_id % 2}
```

Sometimes, your SQL might not contain the sharding key, but the sharding key actually *exists in the external logic*. In this scenario, you can use the `Hint` sharding strategy and code sharding conditions to realize flexible sharding routing. `shardingAlgorithmName` is used to specify the sharding algorithm, while the common configuration of the `Hint` sharding strategy can be broken down into simple steps. The following steps guide you through the configuration.

To get started, you could refer to the following script:

```YAML
rules:
```

```
- !SHARDING
  tables:
    t_order:
      databaseStrategy:
        hint:
          shardingAlgorithmName: database_hint
```

To configure the `hint` algorithm, you could refer to the following script:

```YAML
shardingAlgorithms:
    database_hint:
        type: CLASS_BASED
        props:
            strategy: HINT
            algorithmClassName: xxx
```

After the configuration is complete, you can utilize `HintManager` to set the database sharding key value and the table sharding key value. The following code snippet is an example of `HintManager` usage. `addDatabaseShardingValue` and `addTableShardingValue` are used to specify the sharding key value for the logical table. Additionally, since `HintManager` uses `ThreadLocal` to maintain the shard key value, you should use the `close` method to close it at the end, or use `try with resource` to close it automatically:

```Java
try (HintManager hintManager = HintManager.getInstance();
      Connection connection = dataSource.getConnection();
      PreparedStatement preparedStatement = connection.
prepareStatement(sql)){
    hintManager.addDatabaseShardingValue("t_order", 2);


    try (ResultSet resultSet = preparedStatement.
executeQuery()) {
        while (resultSet.next()) {
        }
    }
}
```

The previous script can help you understand how `HintManager` works.

Distributed transaction

The distributed database solution also supports distributed transactions, including **XA** and **BASE**. Distributed transactions provide eventual consistency semantics. In XA mode, if the isolation level of the storage database is *serializable*, strong consistency semantics can be achieved. The following code snippet showcases how to use XA:

```yaml
YAML
 // Proxy server.yaml
rules:
  - !AUTHORITY
    users:
      - root@%:root
      - sharding@:sharding
    - !TRANSACTION
    defaultType: XA
    providerType: Atomikos
```

To use the **Narayana** implementation, `providerType` will be configured as `narayana` and will add the `narayana` dependencies. The following script is the perfect example to refer to:

```
Plain Text
// shardingsphere/pom.xml Delete these item's scope

<dependency>
      <groupId>org.jboss.narayana.jta</groupId>
      <artifactId>jta</artifactId>
      <version>${narayana.version}</version>
</dependency>
<dependency>
      <groupId>org.jboss.narayana.jts</groupId>
      <artifactId>narayana-jts-integration</artifactId>
      <version>${narayana.version}</version>
</dependency>
<dependency>
      <groupId>org.jboss</groupId>
```

```
    <artifactId>jboss-transaction-spi</artifactId>
    <version>${jboss-transaction-spi.version}</version>
</dependency>
<dependency>
    <groupId>org.jboss.logging</groupId>
    <artifactId>jboss-logging</artifactId>
    <version>${jboss-logging.version}</version>
</dependency>

// shardingsphere/shardingsphere-kernel/shardingsphere-
transaction/shardingsphere-transaction-type/shardingsphere-
transaction-xa/shardingsphere-transaction-xa-core/pom.xml add
dependencies
<dependency>
    <groupId>org.apache.shardingsphere</groupId>
    <artifactId>shardingsphere-transaction-xa-narayana</
artifactId>
    <version>${project.version}</version>
</dependency>
```

You can directly package `narayana` for use by modifying the source code dependencies as shown in the preceding code.

HA and the read/write splitting strategy

The following is the YAML script to set the **writeDataSourceName** and **readDataSourceNames** parameters to specify the writer data source and reader data source:

```
YAML
rules:
- !READWRITE_SPLITTING
  dataSources:
    readwrite_ds:
        writeDataSourceName: write_ds
        readDataSourceNames: read_ds_0,read_ds_1
```

Dynamic read/write splitting is a type of read/write splitting jointly used with HA. HA can discover the primary-secondary relationship of the underlying databases and dynamically correct the connection between Apache ShardingSphere and the databases, further ensuring HA at the application layer. The configuration of dynamic read/write splitting is shown here:

```yaml
YAML
rules:
- !DB_DISCOVERY
  dataSources:
    readwrite_ds:
      dataSourceNames:
        - ds_0
        - ds_1
        - ds_2
      discoveryHeartbeatName: mgr-heartbeat
      discoveryTypeName: mgr
```

As you can see, you do not need to specify the write data source and read data source in the configuration of read/write splitting because HA can automatically detect the writer and the reader.

The following step is to proceed to configure `discoveryHeartbeats` so that the system can know which resources are available for you to configure:

```yaml
YAML
  discoveryHeartbeats:
    mgr-heartbeat:
      props:
        keep-alive-cron: '0/5 * * * * ?'
  discoveryTypes:
    mgr:
      type: MGR
      props:
        group-name: 92504d5b-6dec-11e8-91ea-246e9612aaf1
```

The `discoveryHeartbeats` configuration step is followed by configuring read/write splitting:

```yaml
YAML
- !READWRITE_SPLITTING
  dataSources:
    readwrite_ds:
        autoAwareDataSourceName: readwrite_ds
```

Once done, you will have successfully configured your read/write splitting strategy with Apache ShardingSphere.

Elastic scaling

Elastic scaling usually involves sharding rule changes and so is closely related to data sharding. Moreover, elastic scaling is compatible with other core features, such as read/write splitting and HA. Therefore, when the system needs scale-out, regardless of what the current configuration is, elastic scaling can be enabled.

There are many elastic scaling configurations, for example, some for performance tuning, some for resource usage limitation, and some can be customized through a **serial peripheral interface** (**SPI**).

The common `server.yaml` configuration is as follows:

```yaml
YAML
scaling:
  blockQueueSize: 10000
  workerThread: 40
  clusterAutoSwitchAlgorithm:
      type: IDLE
      props:
          incremental-task-idle-minute-threshold: 30
  dataConsistencyCheckAlgorithm:
      type: DEFAULT
mode:
  type: Cluster
  repository:
      type: ZooKeeper
```

```
    props:
        namespace: governance_ds
        server-lists: localhost:2181
        retryIntervalMilliseconds: 500
        timeToLiveSeconds: 60
        maxRetries: 3
        operationTimeoutMilliseconds: 500
  overwrite: false
```

If you are planning on leveraging elastic scaling, the previous script will become fairly familiar to you.

Distributed governance

Apache ShardingSphere supports three operating modes, namely `Memory`, `Standalone`, and `Cluster`:

- In `Memory` mode, metadata is stored in the current process.
- In `Standalone` mode, metadata is stored in a file.
- For `Cluster` mode, the storage metadata and coordination of the state of each computing node occurs in the registry center.

Apache ShardingSphere integrates **ZooKeeper** and **etcd** internally, which enables it to use event notification of registry node changes to synchronize metadata between clusters and configuration information.

Next, we will show you how to configure the three modes of Apache ShardingSphere. Let's start with how to configure `Memory` mode. The first will be to specify the mode, as follows:

```YAML
mode:
  type: Memory
```

It is very simple to configure `Memory` mode, and as you will see in the following script, the same can be said about `Standalone` mode:

```YAML
mode:
  type: Standalone
```

```
  repository:
    type: File
  overwrite: false
```

The remaining one, Cluster mode, requires a little more coding in the script:

```YAML
mode:
  type: Cluster
  repository:
    type: ZooKeeper
    props:
      namespace: governance_ds
      server-lists: localhost:2181
      retryIntervalMilliseconds: 500
      timeToLiveSeconds: 60
      maxRetries: 3
      operationTimeoutMilliseconds: 500
  overwrite: false
```

Although the script is a little longer, Cluster mode is nevertheless just as easy to configure, as you can see.

Now, we can jump to security-focused features. The next section will guide you through real-world examples for building your database security solution.

Recommended database security solution

In terms of data security, Apache ShardingSphere provides you with a reliable but easy data encryption solution along with authentication and privilege control.

Simply put, the encryption engine and built-in encryption algorithms provided by ShardingSphere can automatically encrypt and store the input information and decrypt it into plaintext when querying, and then send it back to the client. In this way, you won't have to worry about data encryption and decryption to secure data storage.

In the following sections, you will find real-world examples to implement ShardingSphere for your database security, the tools at your disposal, and how to configure them for data encryption, data migration with encryption, authentication, and SQL authority.

Implementing ShardingSphere for database security

The specific architecture diagram of the data security solution is presented in *Figure 11.3*. The solution consists of core components such as authentication, authority checker, encrypt engine, encryption algorithm, and online legacy unencrypted data processor:

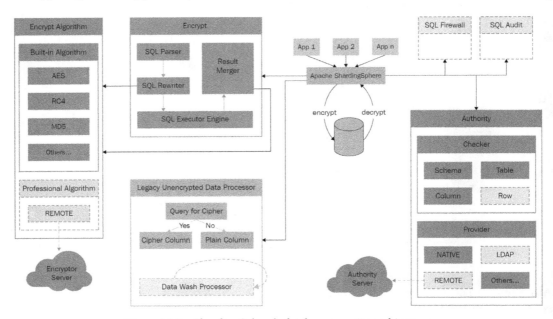

Figure 11.3 – ShardingSphere's database security architecture

As you can see from the previous figure, ShardingSphere can help build a complete solution solving all of your database security concerns. You may be wondering what the key aspects are that you should be mindful of when it comes to database security and ShardingSphere, and the following list will help you answer this question.

Let's now learn more about the key elements of database security:

- **Authentication** – Currently, Apache ShardingSphere supports the password authentication protocols of MySQL (`mysql_native_password`) and PostgreSQL (`md5`). In the future, ShardingSphere will add more authentication methods.

- **Privilege control** – Apache ShardingSphere provides two levels of permission control policies, `ALL_PRIVILEGES_PERMITTED` and `SCHEMA_PRIVILEGES_PERMITTED`. You can choose to use one of them or implement an SPI for a custom extension.

- **Encryption engine** – Based on the encryption rules, the engine can parse the input SQL and automatically calculate and rewrite the content to be encrypted and store the ciphertext in the storage node. When you query encrypted data, ShardingSphere can decrypt and output the plaintext.

- **Encryption algorithm** – ShardingSphere's built-in encryption algorithms contain AES, RC4, and MD5, for example, and you can also customize other extensions through an SPI.

- **Legacy unencrypted data processor** (*under development*) – This function can convert the legacy plaintext data in databases into a ciphertext for storage to help you complete historical data migration and system upgrades.

The next section will guide you through the tools that are available to you for setting up your database security solution.

Two clients to choose from

Both ShardingSphere-JDBC and ShardingSphere-Proxy can access the encryption engine and encryption algorithms, and their performance is exactly the same.

However, centralized authentication and authority control are unique features of ShardingSphere-Proxy. For more details, please refer to the *User authentication* and the *SQL authority* sections in *Chapter 4*, *Key Features and Use Cases – Focusing on Performance and Security*.

In the future, the data processor will also be integrated into ShardingSphere-Proxy.

Applying a data security solution to your DBMS

Similar to the sharding solution, the data security solution is independent of the type of storage node and is implemented on the basis of standard SQL. Therefore, you can apply ShardingSphere to access MySQL, PostgreSQL, or any relational databases complying with the SQL-92 standard, and the user experience will be the same. In the future, the data security feature will support more heterogeneous scenarios, allowing you to meet your requirements.

Data encryption/data masking

Data encryption makes distributed databases even more secure. ShardingSphere frees you from the encryption process but enables you to directly use it. The following is an encryption configuration example:

```YAML
rules:
- !ENCRYPT
  encryptors:
    aes_encryptor:
      type: AES
      props:
        aes-key-value: 123456abc
      tables:
            t_encrypt:
      columns:
        user_id:
          cipherColumn: user_cipher
          encryptorName: aes_encryptor
```

In addition to the AES and MD5 encryption algorithms, ShardingSphere supports other encryption algorithms, such as SM3, SM4, and RC4. You can also load custom algorithms via the SPI. In terms of the configuration of encrypted fields, ShardingSphere also provides configurations of cipher fields, plaintext fields, cipher query assistant fields, and so on. You can choose an appropriate configuration method that fits your actual scenarios.

Data migration with encryption

The previous section described how to enable data encryption. Data encryption is only valid for new or updated data, while read-only legacy data cannot be encrypted. Currently, you would need to encrypt legacy data by yourself. To fix the issue and help improve efficiency, ShardingSphere is going to release the legacy data encryption feature soon.

Categorized by destination, there are two types of data migration with encryption: encrypt data migrated to a new cluster and encrypt data in the original cluster.

To achieve *data migration to a new cluster* with encryption, you need to know the implementation method and the steps. The steps for data migration with encryption are similar to the elastic scaling steps we introduced you to in the previous section. This method takes a relatively long time, so if you only need data encryption, we don't recommend this method. However, when you happen to want elastic scaling, you can consider this method and implement elastic scaling at the same time.

In terms of *data encryption in the original cluster*, only a small amount of data needs migrating, and only the new encryption-related columns need to be processed, which is not time-consuming at all. If you need data encryption, choose this method.

Authentication

We have already outlined how to implement user authentication in a distributed database in *Chapter 4*, *Key Features and Use Cases – Focusing on Performance and Security*. Here is the sample configuration provided by ShardingSphere:

```
YAML
rules:
  - !AUTHORITY
    users:
      - root@%:root
      - sharding@:sharding
```

This configuration shows how to define two different users, that is, root and sharding with the same password and username.

The user login addresses are not recruited, allowing them to connect to ShardingSphere from any host and log in.

When you or the administrator need to restrict a user login address, please refer to the following format:

```
YAML
rules:
  - !AUTHORITY
    users:
      - root@%:root
      - sharding@:sharding
      - user1@127.0.0.1:password1
      - user2@192.168.1.11:password2
```

In this way, `user1` and `user2` can only log in from a given IP and cannot connect to ShardingSphere from other addresses. This is a very important security control measure.

In production applications, it is recommended that users may choose to restrict login addresses if the network environment allows it.

SQL authority/privilege checking

In the *SQL authority* section in *Chapter 4, Key Features and Use Cases – Focusing on Performance and Security*, we talked about the two level-authority providers:

- `ALL_PRIVILEGES_PERMITTED`:
  - All permissions are granted to users without any permission restrictions.
  - Default and no configuration required.
  - Suitable for testing and verification, or an application environment where you are absolutely trusted.

- `SCHEMA_PRIVILEGES_PERMITTED`:
  - Grants user access to the specified schema.
  - Requires specific configuration.
  - Suitable for application environments where the scope of user access needs to be limited.

The administrator needs to choose one based on the real application environment.

The next section allows you to have the best use case reference for synthetic monitoring.

Recommended synthetic monitoring solution

Full-link stress testing is a complex and huge project because it requires the cooperation of various microservices and middleware. Apache ShardingSphere's database solution for full-link online stress testing scenarios is referred to as the **stress testing shadow database** function created to help you isolate stress testing data and avoid contaminating data.

Openness and cooperation have always been natural attributes of the open source community. Apache ShardingSphere, together with Apache APISIX and Apache SkyWalking, jointly launched the project **CyborgFlow**, a low-cost **out-of-the-box (OOTB)** full-link online stress testing solution capable of analyzing data traffic in a full link from a unified perspective.

As you can see from *Figure 11.4*, CyborgFlow is a complete stress testing and monitoring solution:

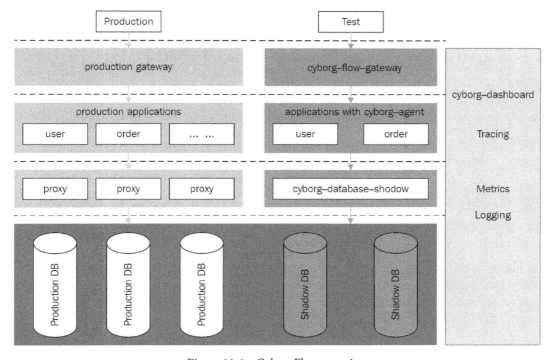

Figure 11.4 – CyborgFlow overview

Now that you have a visual overview of the synthetic monitoring solution, let's jump into the specifics in the following sections.

Flow gateway

To perform online pressure testing in a production environment, we need to enable the stress testing gateway layer to undertake the testing traffic.

The stress testing gateway plugin, cyborg-flow-gateway, is implemented by **Apache APISIX** in accordance with the **SkyWalking** protocol. According to user configuration, the gateway can encapsulate the specified traffic and pass it with the stress testing identifier to the context of the entire link call. The gateway supports authentication and authorization flow processing according to the needs of different services. After completion, users only need to release resources, making the process transparent to the production environment.

Application performance monitoring and Cyborg Agent

With Apache SkyWalking's **Cyborg Dashboard**, you can monitor changes in the stress testing environment from a centralized perspective, which facilitates reasonable intervention in the flow of stress testing and ensures smooth progress.

Cyborg Agent has the ability to transparently transmit stress testing markers across the link. When an application calls the shadow database, `cyborg-database-shadow`, it can intercept the SQL and append the stress testing identifier to it by means of annotations.

In addition, Cyborg Agent leverages the bytecode technology to free users from manually tracking events and the service deployment is stateless.

Database shield

Cyborg database shadow empowered by Apache ShardingSphere can isolate data based on stress testing identifiers.

The shadow database can parse SQL statements and find stress test identifiers in the annotation. Based on the user-configured `HINT` shadow algorithm, it can determine stress test identifiers. If it successfully finds the identifier in the SQL statement, the SQL is routed to the stress testing data source.

This concludes the synthetic monitoring solution section. The next section will introduce the best use cases for the database gateway solution.

Recommended database gateway solution

A **database gateway** is an entry for database cluster traffic. It shields complex connections between applications and database clusters, and therefore, upper-layer applications connected to the database gateway don't need to care about the real state of the underlying database cluster. Additionally, with certain configurations, other features, such as traffic redistribution and traffic governance, can be implemented.

Apache ShardingSphere built above databases can provide enhanced capabilities and manage the data traffic between the application and the database. Thereby, ShardingSphere naturally becomes a database gateway.

Overview and architecture

The overall architecture of the Apache ShardingSphere database gateway solution is shown in the following figure. The core components include read/write splitting and the registry center. Leveraging the distributed governance capabilities of the registry, ShardingSphere can flexibly manage the status and the traffic of computing nodes. The underlying database state of read/write splitting is also maintained in the registry center, and thus by disabling/enabling a read database, you can govern read/write splitting traffic:

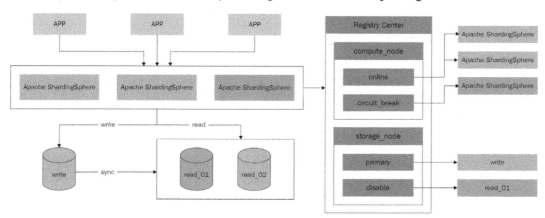

Figure 11.5 – ShardingSphere's database gateway solution

Equipped with the overview provided in *Figure 11.5*, let's now dive into the specifics.

Database management

Database management is a form of ShardingSphere's database traffic governance. Database management consists of two parts:

- Managing ShardingSphere instances
- Managing real database nodes

In fact, database management is based on the distributed governance function of ShardingSphere. The ability provided by Cluster mode can ensure data consistency, state consistency, and service HA of the cluster when deploying online. ShardingSphere implements the distributed governance capabilities of ZooKeeper and etcd by default.

ShardingSphere instances and real database nodes are stored in the registry center as computing nodes and storage nodes, respectively. Their states can be managed through performing operations on the data in the registry center, and then results will be synchronized in real time to all computing nodes in the cluster. The storage structure of the registry center is as follows:

```Bash
namespace
    ├──status
    ├        ├──compute_nodes
    ├        ├        ├──online
    ├        ├        ├        ├──${your_instance_ip_a}@${your_
instance_port_x}
    ├        ├        ├        ├──${your_instance_ip_b}@${your_
instance_port_y}
    ├        ├        ├        ├──....
    ├        ├        ├──circuit_breaker
    ├        ├        ├        ├──${your_instance_ip_c}@${your_
instance_port_v}
    ├        ├        ├        ├──${your_instance_ip_d}@${your_
instance_port_w}
    ├        ├        ├        ├──....
    ├        ├──storage_nodes
    ├        ├        ├──disable
    ├        ├        ├        ├──${schema_1.ds_0}
    ├        ├        ├        ├──${schema_1.ds_1}
    ├        ├        ├        ├──....
    ├        ├        ├──primary
    ├        ├        ├        ├──${schema_2.ds_0}
    ├        ├        ├        ├──${schema_2.ds_1}
    ├        ├        ├        ├──....
```

You can see that ShardingSphere provides the online status and circuit_breaker to manage computing nodes. The latter is used to temporarily shut down the traffic of ShardingSphere instances, and then the instances cannot provide external services until you change the status to online. The availability of both the online status and circuit_breaker makes it convenient to manage the state of computing nodes, that is, ShardingSphere instances:

1. Write DISABLED (case-insensitive) in the IP address @PORT node to break the instance, and delete DISABLED to enable it.

   The ZooKeeper command is as follows:

   ```Bash
   [zk: localhost:2181(CONNECTED) 0] set /${your_zk_
   namespace}/status/compute_nodes/circuit_breaker/${your_
   instance_ip_a}@${your_instance_port_x} DISABLED
   ```

2. Use DistSQL to quickly manage the status of computing nodes:

   ```Bash
   [enable / disable] instance IP=xxx, PORT=xxx
   ```

   Here is an example:

   ```Apache
   disable instance IP=127.0.0.1, PORT=3307
   ```

> **Note**
>
> Although DistSQL has no restrictions on it, it is still not recommended to break the ShardingSphere instances connected to the current client, because the operation may cause the client to become incapable of executing any commands. This issue has been fixed in the updated version 5.1.1 of ShardingSphere.

In terms of storage node management, ShardingSphere provides disable and primary: the former is used for data nodes while the latter is used to manage the primary database.

Storage node management is usually applied for managing the primary database and the secondary databases, also known as the write database and the read database in the read/write splitting scenario.

Similarly, ShardingSphere manages computing nodes to temporarily disable the traffic of a read database. After being disabled, the secondary database cannot allocate any read traffic. No matter which read/write splitting strategy is configured by the user, all `Select` requests from the application are undertaken by other read databases before the data node is re-enabled.

Let's look at an example implementation case. In the read/write splitting scenario, users can write `DISABLED` (case-insensitive) in the data source name's subnode to disable the secondary database's data source, and delete `DISABLED` or the node to enable it. Let's see the necessary steps to configure the read/write splitting scenario:

1. The ZooKeeper command is as follows:

    Bash
    ```
    [zk: localhost:2181(CONNECTED) 0] set /${your_zk_
    namespace}/status/storage_nodes/disable/${your_schema_
    name.your_replica_datasource_name} DISABLED
    ```

2. Use `DistSQL` to quickly manage the storage node status:

    Bash
    ```
    [enable / disable] readwrite_splitting read xxx [from
    schema]
    ```

    Here is an example:

    Bash
    ```
    disable readwrite_splitting read resource_0
    ```

Read/write splitting

Read/write splitting is one of the typical application scenarios of data traffic governance in ShardingSphere. With the increasing access volume of application systems, the challenge to solve the throughput bottleneck is inevitable. The read/write splitting architecture is the mainstream solution at present; by splitting databases into primary databases and secondary databases, the former handles transactional addition, deletion, and modification operations, and the latter processes query operations. This method can effectively avoid row locks caused by data updates, and significantly improve the query performance of the entire system.

By configuring one primary database with multiple secondary databases, query requests can be evenly distributed to multiple data copies to further improve the processing capacity of the system. ShardingSphere supports the *one primary with multiple secondary* architecture well, and its YAML configuration is as follows:

```
YAML
schemaName: readwrite_splitting_db
dataSources:
  primary_ds:
    url: jdbc:postgresql://127.0.0.1:5432/demo_primary_ds
    username: postgres
    password: postgres
  replica_ds_0:
# omitted data source config
  replica_ds_1:
# omitted data source config
```

The rules necessary to complete the configuration are as follows:

```
rules:
- !READWRITE_SPLITTING
  dataSources:
    pr_ds:
      writeDataSourceName: primary_ds
      readDataSourceNames:
        - replica_ds_0
        - replica_ds_1
      loadBalancerName: loadbalancer_pr_ds
```

While the necessary load balancers are as follows:

```
loadBalancers:
    loadbalancer_pr_ds:
      type: ROUND_ROBIN
```

In the preceding example, one write database is configured through `writeDataSourceName`, two read databases are configured through `readDataSourceNames`, and a load balancing algorithm named `loadbalancer_pr_ds` is configured as well. ShardingSphere has two built-in load balancing algorithms that users can configure directly:

- A polling algorithm with the `ROUND_ROBIN` configuration type
- A random algorithm with the `RANDOM` configuration type

ShardingSphere provides users with OOTB built-in algorithms that can meet most user scenarios. If the preceding two algorithms cannot meet your scenarios, you, or developers in general, can also define your own load balancing algorithms by implementing the `ReplicaLoadBalanceAlgorithm` algorithm interface. It's recommended to commit the algorithm to the Apache ShardingSphere community to help other developers with the same needs.

The read/write splitting configuration of ShardingSphere can be used independently or together with data sharding. To ensure consistency, all transactional read/write operations should use the write database.

In addition to custom read-database load balancing algorithms, ShardingSphere also provides `Hint` to force the routing of traffic to the write database.

In real application scenarios, you should pay attention to the following tips:

- ShardingSphere is not responsible for data synchronization between the write database and the read database, and users need to handle it based on the database type.
- Users need to deal with data inconsistency caused by delays in data synchronization of the primary database and the secondary database according to the database type.
- Currently, ShardingSphere does not support multiple primary databases.

This concludes our read/write splitting section. This is a mainstay feature of the ShardingSphere ecosystem, and if you are looking to apply the theory you have picked up for this feature, all you have to do is continue reading.

Summary

If this chapter has achieved the aim we set for it, you should be able to better strategize your ShardingSphere integration with your databases.

Being an ecosystem that has quickly grown to include not only multiple features but multiple clients and deployment possibilities as well, ShardingSphere may seem challenging at first.

While this is absolutely not the case, and while we built this project to simplify database management, the size of the project in itself merits clarification. Thanks to this chapter, we hope that you will be able to better plan your approach to integrating ShardingSphere and make informed decisions.

Nevertheless, at this point, we are still talking about theory, so you may be wondering about the last step of putting theory into practice.

You will find that the next chapter is perfectly tied in with the use cases you have encountered here. Starting with the first section, you will find real-world examples combining multiple features. You can base your future work on these examples, as we believe they will support you in getting things started.

12

Applying Theory to Practical Real-World Examples

In the previous chapter, we went through some useful use cases, developed thanks to ShardingSphere's many years of experience in an enterprise setting.

Based on these same use case experiences, we will now set out to provide you with methodologies to apply the use cases into practical experience.

Providing you with such examples means that by the end of this chapter, you will be able to ground the knowledge you have gained so far on **ShardingSphere** into real-world use cases. In other words, you will be able to apply theory to practice.

Over the course of this chapter, we will be covering the following topics:

- Distributed database solution
- Database security
- Synthetic monitoring
- Database gateway

Technical requirements

No hands-on experience in a specific language is required but it would be beneficial to have some experience in Java since ShardingSphere is coded in Java.

To run the practical examples in this chapter, you will need the following tools:

- **A 2 cores 4 GB machine with Unix or Windows OS**: ShardingSphere can be launched on most OSs.

- **JRE or JDK 8+**: This is the basic environment for all Java applications.

- **Text editor** (**not mandatory**): You can use Vim or VS Code to modify the YAML configuration files.

- **A MySQL/PG client**: You can use the default CLI or other SQL clients such as Navicat or DataGrip to execute SQL queries.

- **7-Zip or tar command**: You can use these tools for Linux or macOS to decompress the proxy artifact.

> **You can find the complete code file here:**
> ```
> https://github.com/PacktPublishing/A-Definitive-
> Guide-to-Apache-ShardingSphere
> ```

Distributed database solution

In order to give you some potential common scenarios that might be encountered in the real world, we have selected a few possible cases. Reading these cases will show you how to combine multiple features to create solutions that can greatly enhance your system, and allow you to make the most of Apache ShardingSphere.

Case 1 – ShardingSphere-Proxy + ShardingSphere-JDBC + PostgreSQL + distributed transaction + cluster mode + the MOD sharding algorithm

The first case will take you through a scenario including ShardingSphere-Proxy as well as ShardingSphere-JDBC working with PostgreSQL. Distributed transactions are included with ShardingSphere's `Cluster` mode and the `MOD` sharding algorithm.

The deployment architecture

The deployment architecture of Case 1 is shown in *Figure 12.1*. Apache ShardingSphere's distributed database solution adopts the hybrid deployment model of ShardingSphere-JDBC plus ShardingSphere-Proxy, and centrally manages sharding rules through a configuration center. The underlying storage engine of the example distributed database is the PostgreSQL database. The **XA transaction manager** manages distributed transactions. The operation mode is Cluster to ensure configuration synchronization between multiple instances. The sharding algorithm is the auto-sharding algorithm, MOD. In such a case, users such as yourself do not need to worry about the underlying data distribution because the automatic sharding algorithm and scaling can help manage shard.

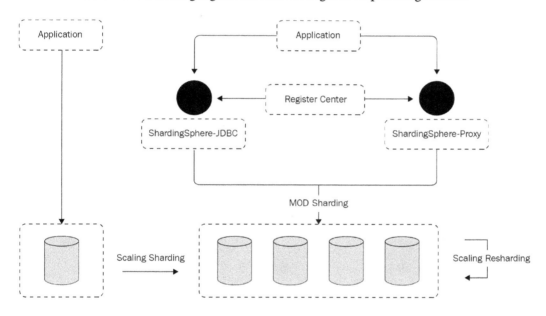

Figure 12.1 – Case 1 deployment architecture

Let's now move on to the example configuration in the next section.

The example configuration

`ShardingSphere-Proxy` `server.yaml` should be configured as follows. The operating mode is `Cluster`, and `ZooKeeper` is the coordinator service used to notify each proxy instance in the cluster to make metadata changes. The XA transaction manager manages transactions, and `Atomikos` is the distributed transaction management solution:

```yaml
YAML
mode:
    type: Cluster
    repository:
        type: ZooKeeper
        props:
    overwrite: true
```

The rules are as follows:

```
rules:
- !AUTHORITY
  users:
    - root@%:root
    - sharding@:sharding
- !TRANSACTION
  defaultType: XA
  providerType: Atomikos
```

The `config-sharding.yaml` configuration file is as follows. The configuration file defines the `sharding_db` logical library and two database resources, `ds_0` and `ds_1`. The sharding rules define the `t_order` table, which uses the `HASH_MOD` automatic sharding algorithm to split the data into four pieces:

```yaml
YAML
schemaName: sharding_db
dataSources:
  ds_0:
#omitted data source config
  ds_1:
#omitted data source config
```

As for the ShardingSphere-JDBC access terminal, the operating mode is configured as `Cluster`, and a unified configuration center is used to manage sharding rules. The metadata of the configuration center is referenced through `schemaName`. The specific configuration is as follows:

```YAML
schemaName: sharding_db

mode:
  type: Cluster
  repository:
    type: ZooKeeper
    props:
      namespace: governance_ds
      server-lists: localhost:2181
  overwrite: false
```

The preceding code completes your distributed database solution preparation for both the proxy and JDBC. The following sections introduce you to starting and testing the configuration.

The recommended cloud/on-premise server

The recommended configuration for the distributed database solution is fairly straightforward. You can refer to the following technical requirements:

- Server configuration:
  - CPU: 8-core
  - Memory: 16 GB
  - Hard disk: 500 GB
- Applications:
  - ShardingSphere-Proxy: 5.0.0
  - PostgreSQL: 14.2
  - ZooKeeper: 3.6.3

Starting and testing the distributed database solution

Follow the next code snippet examples to test your distributed database solution after setting it up, and before starting to use it:

1. Use ShardingSphere-Proxy to create a sharding table rule with `"sharding-count"=4`:

```SQL
psql -U root -d sharding_db -h 127.0.0.1 -p 3307

CREATE SHARDING TABLE RULE t_user (
    RESOURCES(ds_0, ds_1),
    SHARDING_COLUMN=id,TYPE(NAME=MOD,PROPERTIES("shard
ing-count"=4))
);
```

2. Create tables and insert data through ShardingSphere-Proxy:

```SQL
CREATE TABLE `t_user` (
  `id` INT(8) NOT NULL,
  `mobile` CHAR(20) NOT NULL,
  `idcard` VARCHAR(18) NOT NULL,
  PRIMARY KEY (`id`)
);
```

Once you've successfully created the tables, you can insert data with the following code:

```
INSERT INTO t_user (`id`, `mobile`, `idcard`) VALUES
(1,18236***857, 220605******08170),
(2,15686***114, 360222******88804),
-- omitted some values
(12,13983***809, 430204******42092);
```

Now, you are ready to get started in testing to ensure that your configuration is correct. The steps in the next section will guide you through your configuration verification.

Verification testing

Let's get started:

1. Log in to the ds_0 instance and query the results:

```SQL
psql -U root -d demo_ds_0 -h 127.0.0.1 -p 5432
USE demo_ds_0;
SELECT * FROM t_user_0;
SELECT * FROM t_user_2;
```

The following screenshot shows an example of the output you will see on your screen after querying the results:

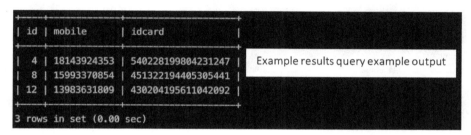

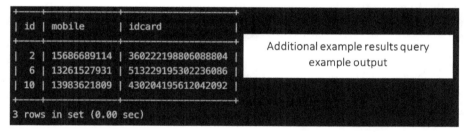

Figure 12.2 – Examples of results query output

2. Log in to the ds_1 instance, query the results, and view the output, as shown in *Figure 12.3*:

```SQL
psql -U root -d demo_ds_1 -h 127.0.0.1 -p 5432
USE demo_ds_1;
SELECT * FROM t_user_1;
SELECT * FROM t_user_3;
```

The outputs are as follows:

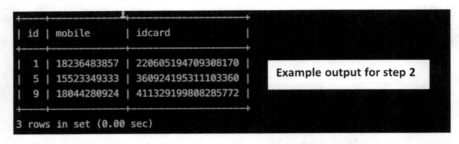

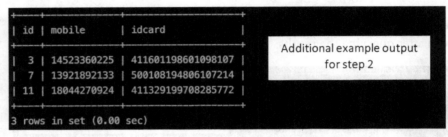

Figure 12.3 – Examples of output for step 2

By executing the query, we can see that the inserted twelve pieces of data are evenly distributed among four shards.

3. Execute RQL statements to validate routing rules and view the outputs, as shown in *Figure 12.4*:

```SQL
SHOW SHARDING TABLE RULES
SHOW SHARDING TABLE NODES
```

The outputs are as follows:

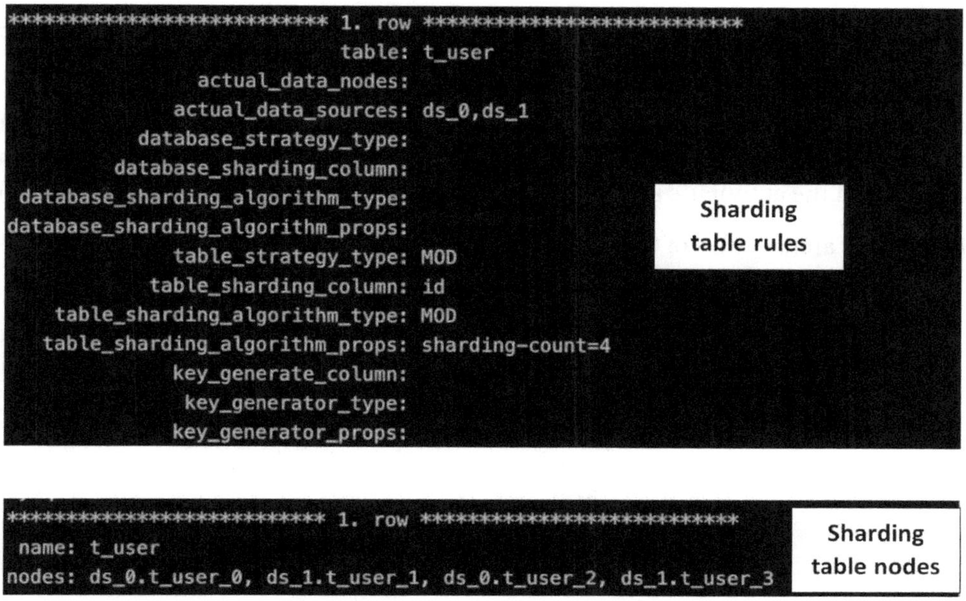

```
*************************** 1. row ***************************
                         table: t_user
             actual_data_nodes:
           actual_data_sources: ds_0,ds_1
         database_strategy_type:
      database_sharding_column:
 database_sharding_algorithm_type:
database_sharding_algorithm_props:
           table_strategy_type: MOD
         table_sharding_column: id
  table_sharding_algorithm_type: MOD
 table_sharding_algorithm_props: sharding-count=4
           key_generate_column:
            key_generator_type:
           key_generator_props:
```

Sharding table rules

```
*************************** 1. row ***************************
  name: t_user
  nodes: ds_0.t_user_0, ds_1.t_user_1, ds_0.t_user_2, ds_1.t_user_3
```

Sharding table nodes

Figure 12.4 – RQL statements example – sharding table rules and nodes

In this section, we have learned how to carry out the verification test for sharding.

Case 2 – ShardingSphere-Proxy + MySQL + read/ write splitting + cluster mode + HA + RANGE sharding algorithm + scaling

The second case we have prepared for you is one that you are bound to encounter if you are considering including Apache ShardingSphere in your system.

The deployment architecture

In terms of Apache ShardingSphere's distributed database solution, ShardingSphere-Proxy is deployed in the Cluster mode, which centrally manages sharding rules through a unified configuration center and synchronizes to multiple ShardingSphere-Proxy instances.

The underlying storage engine of the distributed database is the MySQL database and the **high-availability (HA)** function based on **MySQL MGR** together with the read/write splitting function can implement dynamic read/write splitting to ensure the HA of the distributed database storage engine. The sharding algorithm is the range-based automatic sharding algorithm, BOUNDARY_RANGE. You do not need to worry about the actual data distribution, and the automatic sharding algorithm and scaling can help manage it.

The deployment architecture of Case 2 is shown in *Figure 12.5*:

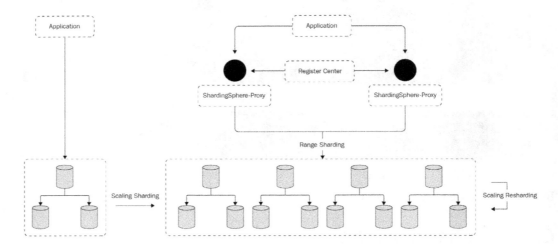

Figure 12.5 – Case 2 deployment architecture

The example configuration

The configuration of ShardingSphere-Proxy server.yaml file is shown as follows. The ShardingSphere operating mode is Cluster mode, and ZooKeeper is the coordinator service used to notify each proxy instance in the cluster to make metadata changes. The transaction manager is the XA transaction manager, while Atomikos provides the distributed transaction management scheme:

```YAML
scaling:
  blockQueueSize: 10000
  workerThread: 40
  clusterAutoSwitchAlgorithm:
    type: IDLE
    props:
```

```
    incremental-task-idle-minute-threshold: 30
dataConsistencyCheckAlgorithm:
    type: DEFAULT
```

One other part of this configuration file is mode:

```
mode:
    type: Cluster
    repository:
        type: ZooKeeper
        props:
            namespace: governance_ds
    overwrite: true
```

Then, we have the rules as follows:

```
rules:
- !AUTHORITY
    users:
        - root@%:root
        - sharding@:sharding
- !TRANSACTION
    defaultType: XA
    providerType: Atomikos
```

The config-sharding-rwsplit-ha.yaml configuration file is as follows. Two sets of MGR clusters are configured in the data source. Auto-discovery of the HA function can automatically identify the primary-secondary relationship. The sharding rule can aggregate two sets of data sources based on the read/write splitting rule and execute the BOUNDARY_RANGE sharding. The sharding algorithm splits the data according to the boundary that you specify. For example, the boundaries 10000000, 20000000, and 30000000 can split the data into four shards.

Now that you understand the components of a ShardingSphere-Proxy server.yaml configuration file, let us understand what a config-sharding-rwsplit-ha.yaml configuration file comprises. Considering the importance that data sharding has within the ShardingSphere ecosystem, as well as its numerous advantages, it is important that you fully master all possible sharding cases.

The `config-sharding-rwsplit-ha.yaml` configuration file is as follows:

```yaml
YAML
schemaName: sharding_db
dataSources:
# omitted datasource config
  primary_ds_0:
  primary_ds_0_replica_0:
  primary_ds_0_replica_1:
  primary_ds_1:
  primary_ds_1_replica_0:
  primary_ds_1_replica_1:
```

Two sets of MGR clusters are configured in the data source. Auto-discovery of the HA function can automatically identify the primary-secondary relationship:

```yaml
- !DB_DISCOVERY
  dataSources:
    pr_ds_0:
      dataSourceNames:
        - primary_ds_0
        - primary_ds_0_replica_0
        - primary_ds_0_replica_1
      discoveryHeartbeatName: mgr-heartbeat
      discoveryTypeName: mgr
```

Then, we have `pr_ds_1`:

```yaml
pr_ds_1:
      dataSourceNames:
        - primary_ds_1
        - primary_ds_1_replica_0
        - primary_ds_1_replica_1
      discoveryHeartbeatName: mgr-heartbeat
      discoveryTypeName: mgr
```

For discovery heartbeats:

```
discoveryHeartbeats:
   mgr-heartbeat:
      props:
         keep-alive-cron: '0/5 * * * * ?'
```

For discovery type:

```
discoveryTypes:
   mgr:
      type: MGR
      props:
         group-name: 92504d5b-6dec-11e8-91ea-246e9612aaf1
```

The sharding rule can aggregate two sets of data sources based on the read/write splitting rule and execute the BOUNDARY_RANGE sharding:

```
!READWRITE_SPLITTING
  dataSources:
    rw_ds_0:
      autoAwareDataSourceName: pr_ds_0
    rw_ds_1:
      autoAwareDataSourceName: pr_ds_1
```

Sharding t_order:

```
- !SHARDING
  tables:
    t_order:
      actualDataNodes: rw_ds_${0..1}.t_order_${0..3}
      tableStrategy:
        standard:
          shardingColumn: order_id
          shardingAlgorithmName: t_order_range
```

Sharding `t_order_item`:

```
t_order_item:
      actualDataNodes: rw_ds_${0..1}.t_order_item_${0..3}
      tableStrategy:
        standard:
          shardingColumn: order_id
          shardingAlgorithmName: t_order_item_range
```

Sharding the algorithms database:

```
shardingAlgorithms:
    database_inline:
      type: INLINE
      props:
        algorithm-expression: rw_ds_${user_id % 2}
```

Sharding the algorithms table:

```
t_order_range:
      type: BOUNDARY_RANGE
      props:
        sharding-ranges: 10000000,20000000,30000000
    t_order_item_range:
      type: BOUNDARY_RANGE
      props:
        sharding-ranges: 10000000,20000000,30000000
```

After completing the preceding configuration and starting `ShardingSphere-Proxy`, you can trigger the scaling job through the following steps:

1.   Add a new database instance and add a new data source in ShardingSphere.
2.   Modify the table sharding rules.

After triggering the scaling job, you can manage the elastic scaling process through DistSQL. If you're using an automated process, just check the job's progress and wait until it's complete.

For the detailed process, please refer to the official documentation of ShardingSphere.

The recommended cloud/on-premises server

The recommended configuration for this case is simple. You can refer to the following technical requirements:

- Server configuration:
  - CPU – 8 Cores
  - Memory – 16 GB
  - Disk – 500 GB

- Application information:
  - ShardingSphere-Proxy 5.0.0
  - MySQL 8.0
  - JDBC-Demo
  - Zookeeper 3.6.3

Starting and testing your case configuration

Follow the next code snippets as an example to test the configuration for the second case we have prepared for you:

1. When `ShardingSphere-Proxy` is started, log in to `ShardingSphere-Proxy` to create a table and insert the following data:

```SQL
mysql -uroot -h127.0.0.1 -P3307 -proot
use sharding_db;
DROP TABLE IF EXISTS t_user;
CREATE TABLE `t_user` (
  `id` int(8) not null,
  `mobile` char(20) NOT NULL,
  `idcard` varchar(18) NOT NULL,
  PRIMARY KEY (`id`)
);
INSERT INTO t_user (id, mobile, idcard) VALUES
(1,18236***857, 220605******308170),
(2,15686***114, 360222******088804),
```

```
(3,14523***225,  411601*****098107),
(4,18143***353,  540228*****231247),
(5,15523***333,  360924*****103360),
(6,13261***931,  513229*****236086),
(7,13921***133,  500108*****107214),
(8,15993***854,  451322*****305441),
(9,18044***924,  411329*****285772),
(10,1398***1809, 430203*****042092);
```

2. Start the JDBC program to deliver query results in real time, as shown in the following screenshot:

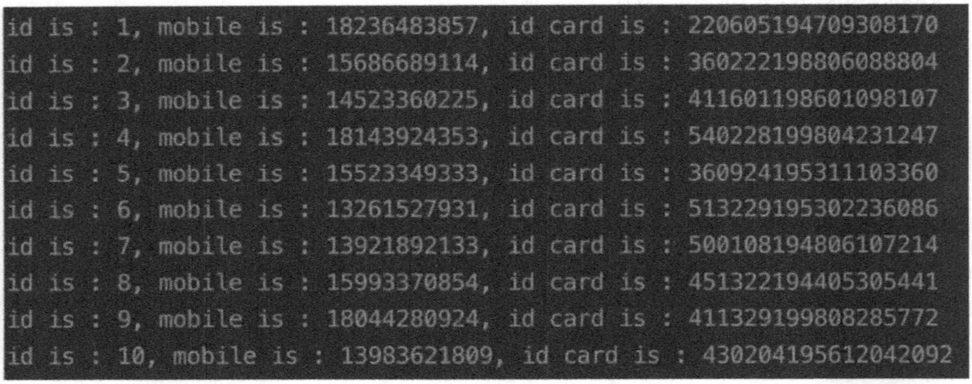

```
id is : 1, mobile is : 18236483857, id card is : 2206051947093081700
id is : 2, mobile is : 15686689114, id card is : 3602221988060888044
id is : 3, mobile is : 14523360225, id card is : 4116011986010981070
id is : 4, mobile is : 18143924353, id card is : 5402281998042312477
id is : 5, mobile is : 15523349333, id card is : 3609241953111033600
id is : 6, mobile is : 13261527931, id card is : 5132291953022360866
id is : 7, mobile is : 13921892133, id card is : 5001081948061072144
id is : 8, mobile is : 15993370854, id card is : 4513221944053054411
id is : 9, mobile is : 18044280924, id card is : 4113291998082857722
id is : 10, mobile is : 13983621809, id card is : 4302041956120420922
```

Figure 12.6 – JDBC query results example

3. View the current relationship between the active and standby nodes through DistSQL:

```
SQL
SHOW DB_DISCOVERY RULES\G


-- View the standby node state
SHOW READWRITE_SPLITTING READ RESOURCES;
```

Figure 12.7 shows you an example output result when querying the system for the current database discovery rules and read/write splitting read resources:

```
mysql> SHOW DB_DISCOVERY RULES\G
*************************** 1. row ***************************
              group_name: primary_replica_ds
        data_source_names: ds_0,ds_1,ds_2
primary_data_source_name: ds_0
          discovery_type: {group_name=mgr, type=mgr, props={group-name=b13df29e-90b6-11e8-8d1b-525400fc3996}}
     discovery_heartbeat: {group_name=mgr-hearbeat, props={keep-alive-cron=0/5 * * * * ?}}
1 row in set (0.07 sec)

mysql> SHOW READWRITE_SPLITTING READ RESOURCES\G
*************************** 1. row ***************************
resource: ds_1
  status: enable
*************************** 2. row ***************************
resource: ds_2
  status: enable
2 rows in set (0.02 sec)
```

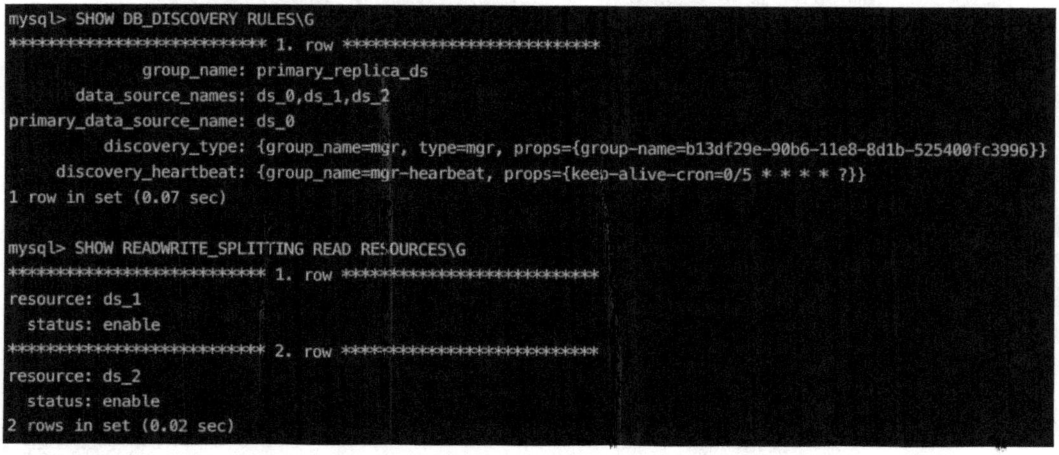

Figure 12.7 – Database discovery and read/write splitting read resources

Shut down the secondary nodes of one set of MGR:

```SQL
mysql -uroot -hprimary_ds_0_replica_0.db -P3306 -p

SHUTDOWN;
```

4. When viewing the node status through proxy, the status of the secondary node that is shut down is disabled:

```SQL
SHOW READWRITE_SPLITTING READ RESOURCES;
```

Once you query your system to show the read resources, you will receive a result as shown in the following figure:

```
mysql> SHOW DB_DISCOVERY RULES\G
*************************** 1. row ***************************
            group_name: primary_replica_ds
      data_source_names: ds_0,ds_1,ds_2
primary_data_source_name: ds_0
         discovery_type: {group_name=mgr, type=mgr, props={group-name=b13df29e-90b6-11e8-8d1b-525400fc3996}}
    discovery_heartbeat: {group_name=mgr-hearbeat, props={keep-alive-cron=0/5 * * * * ?}}
1 row in set (0.00 sec)

mysql>
mysql>
mysql> SHOW READWRITE_SPLITTING READ RESOURCES\G
*************************** 1. row ***************************
resource: ds_2
  status: enable
*************************** 2. row ***************************
resource: ds_1
  status: disable
2 rows in set (0.01 sec)
```

Figure 12.8 – Read/write splitting example output

5. JDBC still queries and delivers the results in real time, as you can see from the screenshot in the following figure:

```
X  App-Demo (ssh)
id is : 2, mobile is : 15686689114, id card is : 360222198806088804
id is : 3, mobile is : 14523360225, id card is : 411601198601098107
id is : 4, mobile is : 18143924353, id card is : 540228199804231247
id is : 5, mobile is : 15523349333, id card is : 360924195311103360
id is : 6, mobile is : 13261527931, id card is : 513229195302236086
id is : 7, mobile is : 13921892133, id card is : 500108194806107214
id is : 8, mobile is : 15993370854, id card is : 451322194405305441
id is : 9, mobile is : 18044280924, id card is : 411329199808285772
id is : 10, mobile is : 13983621809, id card is : 430204195612042092
```

Figure 12.9 – JDBC real-time query results

6. To shut down the primary node of MGR, we will first insert some data through proxy:

```SQL
INSERT INTO t_user (id, mobile, idcard) VALUES
```

```
(11,1392***2134,  500108******07211),
(12,1599***0855,  451322******05442),
(13,1804***0926,  411329******85773),
(14,1398***1807,  430204******42094),
(15,1804***0928,  411329******85775),
(16,1398***1800,  130204******42096),
(17,1398***1800,  230204******42093),
(18,1398***1800,  330204******42091),
(19,1398***1800,  230204******42095),
(20,1398***1811,  230204******42092);
```

Then, shut down the primary node:

```
SQL
mysql -uroot -hprimary_ds_0.db -P3306 -p
SHUTDOWN;
```

We can tell from the JDBC program that when the primary node SHUTDOWN is queried later and affected by the SQL insertion of the previous code block, the newly inserted data is also queried. As you can see from the screenshot in the following figure, the SQL insertion that the data has successfully been inserted into is t_user:

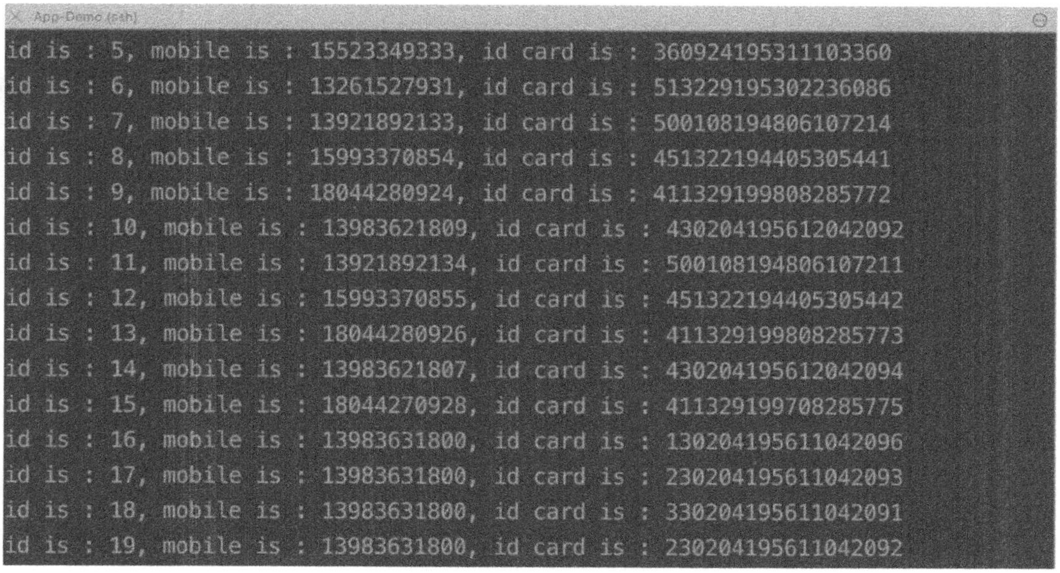

Figure 12.10 – SQL insertion screenshot

This is a test of applicable HA scenarios.

Database security

After covering some distributed database real-world examples that will come in handy, let's now move on to some database security examples.

Case 3 – ShardingSphere-Proxy + ShardingSphere-JDBC + PostgreSQL + data encryption

In this section, the first case will get you started with data encryption, while the following case will cover checking privileges.

The deployment architecture

This case demonstrates how to use ShardingSphere-Proxy to dynamically manage rule configuration, connecting PostgreSQL to ShardingSphere-JDBC and implementing a data encryption application.

To understand what the deployment would look like, you can refer to *Figure 12.11*:

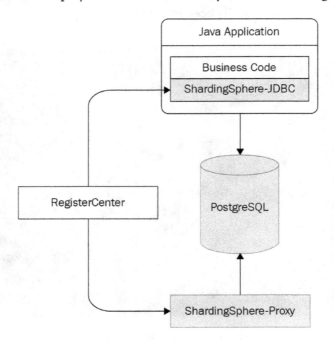

Figure 12.11 – Case 3 deployment architecture

The example configuration

For this case, you can use either ShardingSphere-Proxy or JDBC, and therefore, in the following sections, you will find an example configuration for both clients.

ShardingSphere-Proxy

The operation mode of `ShardingSphere-Proxy` is configured as `Cluster`, and the content of `server.yaml` is as follows:

```yaml
YAML
mode:
  type: Cluster
  repository:
    type: ZooKeeper
    props:
      namespace: governance_ds
      server-lists: localhost:2181
  overwrite: true
rules:
- !AUTHORITY
  users:
    - root@%:root
```

Use `config-encrypt.yaml` to create a logic database and add the storage resource as in the following script:

```yaml
YAML
schemaName: encrypt_db
dataSources:
  ds_0:
    url: jdbc:postgresql://127.0.0.1:5432/demo_ds_0
    username: postgres
    password: postgres
```

When `ShardingSphere-Proxy` is started, connect to encrypt_db using the `psql` command and create the required encryption rules through DistSQL:

```sql
SQL
CREATE ENCRYPT RULE t_encrypt (
COLUMNS (
```

```
(NAME=password,CIPHER=password_cipher,TYPE(NAME=AES,PROPERTIES(
'aes-key-value'='123456abc'))))
);
```

As such, the configuration of proxy is done.

ShardingSphere-JDBC

Since unified configuration through ShardingSphere-Proxy is done, we only need to configure the information of the governance center when connecting to ShardingSphere-JDBC applications. Here is an example:

```
YAML
schemaName: encrypt_db
mode:
  type: Cluster
  repository:
    type: ZooKeeper
    props:
      namespace: governance_ds
      server-lists: localhost:2181
  overwrite: false
```

Great, the configuration is done! After JDBC is started, we can also dynamically add or modify encryption rules by executing DistSQL through proxy, avoiding the need to restart the application when modifying its configuration, and providing a flexible management solution.

The recommended cloud/on-premises server

The following list presents you with an example server configuration and application information:

- Server configuration:

  - CPU: 8-core

  - Memory: 16 GB

  - Disk: 500 GB

- Application information:

  - ShardingSphere: 5.0.0

- PostgreSQL: 14.2
- ZooKeeper: 3.6.3

Starting and testing your database security solution

Testing your secure database solution will involve a four-step procedure. You can find it with the following code:

1. Start `ShardingSphere-Proxy` and execute the prepared DistSQL.

2. View the resource and encryption rules on the proxy side:

    ```SQL
    SHOW SCHEMA RESOURCES FROM encrypt_db;
    SHOW ENCRYPT RULES FROM encrypt_db;
    ```

3. Execute **data definition language** (**DDL**) on the proxy side to create a data table:

    ```SQL
    DROP TABLE IF EXISTS t_encrypt;
    CREATE TABLE t_encrypt (
        id int NOT NULL,
        name varchar DEFAULT NULL,
        password varchar DEFAULT NULL,
        PRIMARY KEY (id)
    );
    ```

4. Start and connect to ShardingSphere-JDBC to read and write data and observe the results of data encryption and decryption.

Successfully completing the fourth step will conclude your testing for your secure database solution. The next section will guide you through more security-related testing.

Case 4 – ShardingSphere-Proxy + MySQL + data masking + authentication + checking privileges

We follow the same pattern here by first introducing you to the deployment architecture, then an example configuration, and finally the testing procedure.

The deployment architecture

This case shows you how to create two different data encryption schemas and how to entitle their access to different users. As you can see in *Figure 12.12*, we use ShardingSphere-Proxy and MySQL for data masking, authentication, and checking privileges:

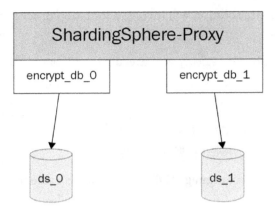

Figure 12.12 – Case 4 deployment architecture

The example configuration

In this example, we use YAML to configure resource and data encryption rules. The configuration is as follows:

- For data source 0 (`ds_0`), let's look at the `config-encrypt-0.yaml` file and its components:

```yaml
YAML
schemaName: encrypt_db_0
dataSources:
 ds_0:
   url: jdbc:mysql://127.0.0.1:3306/demo_
ds_0?useSSL=false
   username: root
   password: 123456
rules:
- !ENCRYPT
  tables:
    t_encrypt_0:
      columns:
```

```
        password:
           cipherColumn: password_cipher
           encryptorName: password_encryptor
     encryptors:
       password_encryptor:
         type: AES
         props:
           aes-key-value: 123456abc
```

The file will have encryptors as shown in the following snippet:

```
encryptors:
    password_encryptor:
      type: AES
      props:
        aes-key-value: 123456abc
```

* For data source 1 (ds_1), let's look at the config-encrypt-1.yaml file and its components as well:

```
YAML
schemaName: encrypt_db_1
dataSources:
  ds_1:
    url: jdbc:mysql://127.0.0.1:3306/demo_
ds_1?useSSL=false
    username: root
    password: 123456
rules:
- !ENCRYPT
  tables:
    t_encrypt_1:
      columns:
        password:
           cipherColumn: password_cipher
           encryptorName: password_encryptor
  encryptors:
    password_encryptor:
      type: AES
```

```
      props:
        aes-key-value: 123456abc
```

It will have rules, as shown in the following code:

```
rules:
- !ENCRYPT
  tables:
    t_encrypt_1:
      columns:
        password:
          cipherColumn: password_cipher
          encryptorName: password_encryptor
  encryptors:
    password_encryptor:
      type: AES
      props:
        aes-key-value: 123456abc
```

And it will have encryptors as well:

```
encryptors:
    password_encryptor:
      type: AES
      props:
        aes-key-value: 123456abc
```

Three users are defined in this configuration, among which, for our example, `user0` and `user1` are banned from logging in as hosts and are given different schema accesses:

```
YAML
rules:
- !AUTHORITY
  users:
    - root@%:root
    - user0@127.0.0.1:password0
    - user1@127.0.0.1:password1
  props:
    user-schema-mappings: root@%=encrypt_db_0,root@%=encrypt_
db_1,user0@127.0.0.1=encrypt_db_0,user1@127.0.0.1=encrypt_db_1
```

According to the rule, user0 can only access logic database encrypt_db_0 when connecting from 127.0.0.1. Similarly, user1 can only access encrypt_db_1 when connecting from 127.0.0.1. Otherwise, they cannot see any database. On the contrary, the root user is not limited by the host and has access to both encrypt_db_0 and encrypt_db_1.

The recommended cloud/on-premises server

You can refer to the following list for an example of server configuration and application information:

- Server configuration:

  - CPU: 8-core

  - Memory: 16 GB

  - Disk: 500 GB

- Application information:

  - ShardingSphere: 5.0.0

  - MySQL: 8.0

Start and test it!

Thanks to ShardingSphere-Proxy, you will find that the procedure for testing your authentication and encryption is very simple. You can follow the next steps as an example and you will find that you can repurpose them in your environment:

1. Start ShardingSphere-Proxy.

2. Connect proxy by root user and view the resource and encryption rules:

```SQL
SHOW SCHEMA RESOURCES FROM encrypt_db_0;
SHOW ENCRYPT RULES FROM encrypt_db_0;
SHOW SCHEMA RESOURCES FROM encrypt_db_1;
SHOW ENCRYPT RULES FROM encrypt_db_1;
```

3. Use the root user to execute DDL, and create the t_encrypt_0 data table in encrypt_db_0:

```SQL
USE encrypt_db_0;
```

```
DROP TABLE IF EXISTS t_encrypt_0;
CREATE TABLE t_encrypt_0 (
    `id` int(11) NOT NULL,
    `name` varchar(32) DEFAULT NULL,
    `password` varchar(64) DEFAULT NULL,
    PRIMARY KEY (`id`)
) ENGINE=InnoDB DEFAULT CHARSET=utf8mb4;
```

In the same way, create the t_encrypt_1 data table in encrypt_db_1:

```
SQL
USE encrypt_db_1;
DROP TABLE IF EXISTS t_encrypt_1;
CREATE TABLE t_encrypt_1 (
    `id` int(11) NOT NULL,
    `name` varchar(32) DEFAULT NULL,
    `password` varchar(64) DEFAULT NULL,
    PRIMARY KEY (`id`)
) ENGINE=InnoDB DEFAULT CHARSET=utf8mb4;
```

4. Use user0 to log into proxy to try different logic databases and see whether the result is up to expectations:

```
SQL
USE encrypt_db_0; # succeed
USE encrypt_db_1; # fail
```

Use user1 to log into proxy and perform the same test:

```
SQL
USE encrypt_db_0; # fail
USE encrypt_db_1; # succeed
```

5. The test is done. Now, user0 and user1 can use their respective logic database to read and write.

This concludes our cases on database security testing. We will now move on to testing your synthetic monitoring.

Synthetic monitoring

Having covered both distributed database and database security cases, we thought you might need a good example on how to implement synthetic monitoring. This would help you have a clear overview of how your system is running. We provide you with one case here, as it is easily generalizable to cover all possible scenarios.

Case 5 – Synthetic monitoring

If you are interested in the synthetic monitoring feature we introduced in *Chapter 4, Key Features and Use Cases – Focusing on Performance and Security*, you can refer to the following sections for a complete guide to be able to test your own solution.

The deployment architecture

For our synthetic monitoring case, let's consider a scenario where you can deploy ShardingSphere-JDBC or ShardingSphere-Proxy alternatively. The underlying storage is PostgreSQL, with a database gateway, a default test strategy, and tracing visualization features. *Figure 12.13* presents you with the example deployment architecture:

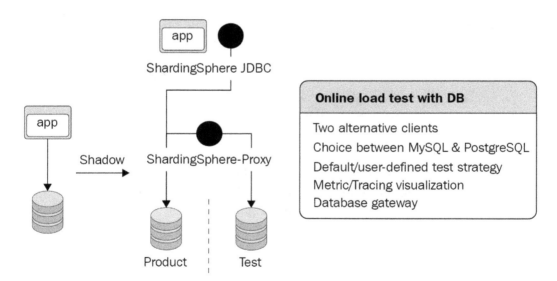

Figure 12.13 – Case 5 deployment architecture

Now that you have visualized the deployment architecture thanks to the previous figure, let's jump into the configuration steps.

The example configuration

As an out-of-the-box, full-link, and online stress testing solution, **CyborgFlow** can be quickly integrated into your projects. Currently, version 0.1.0 has been released.

Here is an example of a `cyborg-database-shadow` configuration showing its components:

1. The following code shows the `server.yaml` configuration:

```YAML
rules:
  - !AUTHORITY
    users:
      - root@%:root
      - sharding@:sharding
  props:
  sql-comment-parse-enabled: true
```

2. Here is the `config-shadow.yaml` configuration:

```YAML
schemaName: cyborg-database-shadow
dataSources:
  ds:
    url: jdbc:mysql://127.0.0.1:3306/
ds?serverTimezone=UTC&useSSL=false
    username: root
    password:
    connectionTimeoutMilliseconds: 30000
    idleTimeoutMilliseconds: 60000
    maxLifetimeMilliseconds: 1800000
    maxPoolSize: 50
    minPoolSize: 1
  ds_shadow:
    url: jdbc:mysql://127.0.0.1:3306/ds_
shadow?serverTimezone=UTC&useSSL=false
    username: root
    password:
    connectionTimeoutMilliseconds: 30000
```

```
    idleTimeoutMilliseconds: 60000
    maxLifetimeMilliseconds: 1800000
    maxPoolSize: 50
    minPoolSize: 1
```

We also have the following rules:

```
rules:
- !SHADOW
  enable: true
  dataSources:
    shadowDataSource:
      sourceDataSourceName: ds
      shadowDataSourceName: ds_shadow
  defaultShadowAlgorithmName: simple-note-algorithm
  shadowAlgorithms:
    simple-note-algorithm:
      type: SIMPLE_NOTE
      props:
        cyborg-flow: true
```

Another example of the cyborg-flow-gateway configuration is as follows:

1. To start, you should note that we use config.yaml here:

```
YAML
apisix:
  config_center: yaml
  enable_admin: false
plugins:
  - proxy-rewrite
  - skywalking
plugin_attr:
  skywalking:
    service_name: APISIX
    service_instance_name: "cyborg-dashboard"
    endpoint_addr: http://127.0.0.1:12800
```

2. Let's see the configuration for `apisix.yaml`:

```yaml
YAML
routes:
  -
    uri: /order
    plugins:
      proxy-rewrite:
        headers:
          sw8-correlation: Y3lib3JnLWZsb3c=:dHJ1ZQ==
      skywalking:
        sample_ratio: 1
    upstream:
      nodes:
        "httpbin.org:80": 1
      type: roundrobin
```

In this example, `proxy-rewrite add-on` was used to inject `sw8-correlation: Y3lib3JnLWZsb3c=:dHJ1ZQ==` into the requested headers. `Y3lib3JnLWZsb3c=` is the Base64 encoding of `cyborg-flow`, and `dHJ1ZQ==` is the Base64 encoding of `true`.

The recommended cloud/on-premises server

In the following list, we provide you with a recommended configuration for cases, whether you are interested in `cyborg-database-shadow` or `cyborg-dashboard`:

- `cyborg-database-shadow`:

  - CPU – 48-core

  - Memory – 96 GB

  - Disk – SSD 820 GB

  - `cyborg-database-shadow` – 0.1.0

  - MySQL – 8.0

  - JDBC-Demo – N/A

  - ZooKeeper – 3.6.3

- `cyborg-dashboard:`

  - CPU – 8-core

  - Memory – 16 GB

  - Disk – SSD 40 GB

  - `cyborg-dashboard` – 0.1.0

  - `cyborg-flow-gateway`

  - CPU – 8-core

  - Memory – 16 GB

  - Disk – SSD 40 GB

  - `cyborg-flow-gateway` – 0.1.0

Starting and testing it

To start, quickly deploy CyborgFlow in a CentOS 7 environment as introduced in *Chapter 4, Key Features and Use Cases – Focusing on Performance and Security*, and activate the default stress testing marker in `/*cyborg-flow: true*/`.

Download related components from `https://github.com/SphereEx/cyborg-flow/releases/tag/v0.1.0`.

Let's first start by preparing the shadow database. To do so, you can refer to the following steps:

1. Prepare the shadow database:

    I.   Unzip `cyborg-database-shadow.tar.gz`.

    II. Configure the business database and shadow database in the `conf/config-shadow.yaml` file as the configuration of your business.

    III. Start the `cyborg-database-shadow` service. The start script is in `bin/start.sh`.

2. Now that you have prepared the shadow database, let's now deploy `cyborg-dashboard`.

    Unzip `cyborg-dashboard.tar.gz`. Start the `cyborg-dashboard` backend and UI interface service for link data monitoring. The start script is in `bin/startup.sh`.

3. Once the dashboard is successfully set up, you can now deploy the stress testing application by `cyborg-agent` with the following three steps:

    I. Unzip `cyborg-agent.tar.gz`.

    II. Modify `collector.backend_service` in `config/agent.config` as the backend address (default access: `11800`) in `cyborg-dashboard` for reporting monitoring data to `cyborg-dashboard`.

    III. When starting the program, add the `-jar {path}/to/cyborg-agent/skywalking-agent.jar` parameter to the `start` demand.

Deploying cyborg-flow-gateway

After completion, you can finally deploy `cyborg-flow-gateway`, as follows:

1. Install OpenResty and Apache APISIX's RPM package:

    ```Bash
    sudo yum install -y https://repos.apiseven.com/packages/
    centos/apache-apisix-repo-1.0-1.noarch.rpm
    ```

2. Install Apache APISIX and all dependencies through the RPM package:

    ```Bash
    sudo yum install -y https://repos.apiseven.com/packages/
    centos/7/x86_64/apisix-2.10.1-0.el7.x86_64.rpm
    ```

3. Modify the `conf/config.yaml` configuration as in the example we provided earlier, which you can find here:

    ```YAML
    apisix:
      config_center: yaml
      enable_admin: false
    plugins:
      - proxy-rewrite
      - skywalking
    plugin_attr:
      skywalking:
        service_name: APISIX
        service_instance_name: "cyborg-dashboard"
        endpoint_addr: http://127.0.0.1:12800
    ```

4. Modify the `conf/apisix.yaml` configuration as in the example we provided earlier in this chapter, which you can find here:

```yaml
YAML
routes:
  -
    uri: /order
    plugins:
      proxy-rewrite:
        headers:
          sw8-correlation: Y3lib3JnLWZsb3c=:dHJ1ZQ==
      skywalking:
        sample_ratio: 1
    upstream:
      nodes:
        "httpbin.org:80": 1
      type: roundrobin
```

Let's assume that an e-commerce website needs to perform online stress testing for its order placement business. Let's also assume that the table related to stress testing is the t_order order table and the test user ID is 0. The data generated by the test user is executed to the ds_shadow shadow database while the production data is executed to the production database, ds.

Let's look at the order table:

```sql
SQL
CREATE TABLE `t_order` (
    `id` INT(11) AUTO_INCREMENT,
    `user_id` VARCHAR(32) NOT NULL,
    `sku` VARCHAR(32) NOT NULL,
    PRIMARY KEY (`id`)
) ENGINE = InnoDB;
```

You can use **Postman** to simulate a request.

5. Directly request the order service to simulate production data:

http://192.168.87.4:8082/order/create

Figure 12.14 – Simulation request using Postman

6. You can view the request link and execute it on `cyborg-dashboard`:

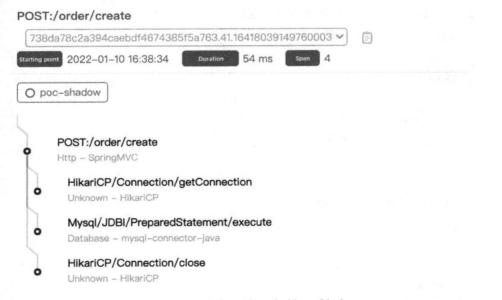

Figure 12.15 – CyborgFlow dashboard link

7. View the data in the production database with the following code:

```Bash
mysql> select * from t_order;
+----+-----------+---------------+
| id | user_id   | sku           |
+----+-----------+---------------+
|  1 | 1         | suk-1-pro     |
+----+-----------+---------------+
1 rows in set (0.00 sec)
```

As you can see from the resulting output, the production data is present with suk-1-pro.

8. Request the cyborg-flow-gateway gateway service to simulate testing data:

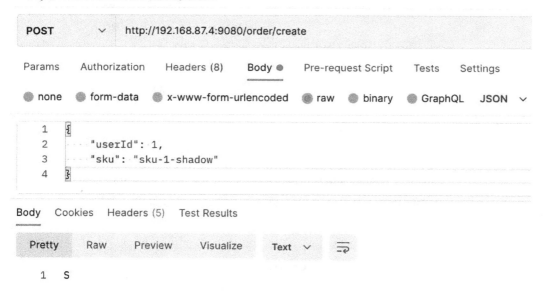

Figure 12.16 – Testing data simulation request in CyborgFlow

9. View the request link and execute SQL in `cyborg-dashboard`:

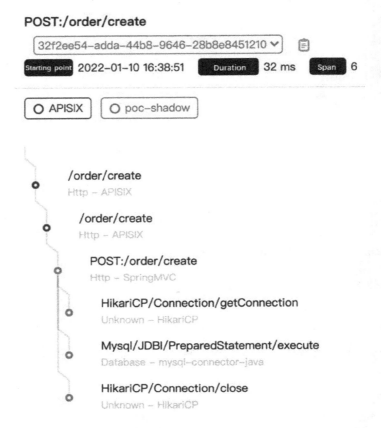

Figure 12.17 – Simulation link and request data test in CyborgFlow

10. View the test data in the `cyborg-database-shadow` database with the following code example:

```Bash
mysql> select * from t_order;

+----+-----------+---------------+
| id | user_id   | sku           |
+----+-----------+---------------+
| 1  | 1         | suk-1-shadow  |
+----+-----------+---------------+
1 rows in set (0.00 sec)
```

As you can see from the output result of the query, the data is present with `suk-1-shadow`. The next section will introduce examples including ShardingSphere's database gateway feature.

Database gateway

In this section, we will go through real real-world examples featuring database gateway. We will first start by reviewing the deployment architecture, followed by the configuration and testing for both of ShardingSphere's clients – Proxy and JDBC.

The deployment architecture

In this case, the underlying database adopts PostgreSQL, deployed with the read/write splitting architecture of *one host and two secondaries*. The upper database layer adopts a hybrid deployment solution of ShardingSphere-Proxy and ShardingSphere-JDBC. Based on the distributed governing capabilities provided by `Cluster` mode, you can easily modify cluster metadata online and synchronize the data to proxy and JDBC. Proxy can use the ShardingSphere built-in DistSQL to achieve traffic control operations (including circuit breaker) and disable the secondary database.

Figure 12.18 illustrates the typical deployment architecture, including both proxy and JDBC, and primary and secondary databases.

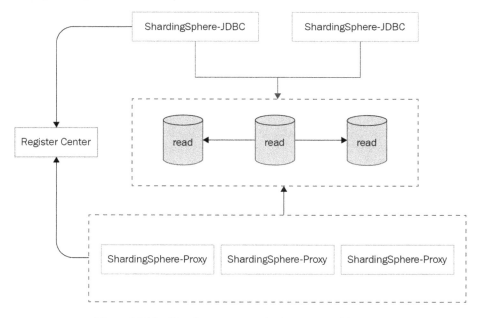

Figure 12.18 – Database gateway deployment architecture

The example configuration

ShardingSphere-Proxy and ShardingSphere-JDBC have to be configured respectively. In actual application scenarios, you can achieve dynamic configuration with DistSQL. This case adopts the YAML configuration to make the configuration easier to understand.

We will first look at how to configure the database gateway with ShardingSphere-Proxy:

1. We first choose the `Cluster` mode in `server.yaml`. Proxies in the same cluster share the same configurations:

```YAML
mode:
  type: Cluster
  repository:
    type: ZooKeeper
    props:
      namespace: governance_ds
      server-lists: localhost:2181
      retryIntervalMilliseconds: 500
      timeToLiveSeconds: 60
      maxRetries: 3
      operationTimeoutMilliseconds: 500
  overwrite: true
props:
  sql-show: true
```

2. Then, configure the read/write splitting rules in `config-readwrite-splitting.yaml`. Proxies in the same cluster need to be configured only once since other proxies will synchronize the configuration from the registry center when initiated:

```YAML
schemaName: readwrite_splitting_db
dataSources: # omitted data source config
  primary_ds:
  replica_ds_0:
  replica_ds_1:
```

3. With the data source, we should also input the rules for the implementation, as presented in the following code:

```
rules:
- !READWRITE_SPLITTING
  dataSources:
    pr_ds:
      writeDataSourceName: primary_ds
      readDataSourceNames:
        - replica_ds_0
        - replica_ds_1
```

We can now look at how to configure the database gateway with ShardingSphere-JDBC. ShardingSphere-JDBC should keep the same `Cluster` mode as ShardingSphere-Proxy. It doesn't require a local configuration of rules but will synchronize the ShardingSphere-Proxy configuration from the registry center when initiated:

```
YAML
schemaName: readwrite_splitting_db
mode:
  type: Cluster
  repository:
    type: ZooKeeper
    props:
      namespace: governance_ds
      server-lists: localhost:2181
      retryIntervalMilliseconds: 500
      timeToLiveSeconds: 60
      maxRetries: 3
      operationTimeoutMilliseconds: 500
  overwrite: false
```

ShardingSphere-JDBC should configure `schemaName` to ensure the same cluster as ShardingSphere-Proxy.

The recommended cloud/on-premises server

You can refer to the following list for an example of server configuration and application information:

- Server configuration:
  - CPU – 8-core
  - Storage – 16 GB
  - Hard Drive – 500 GB
- Application information:
  - ShardingSphere-Proxy – 5.0.0
  - PostgreSQL – 14.2
  - ZooKeeper – 3.6.3

Start and test it!

If you have followed the preceding steps to configure your database gateway with both ShardingSphere Proxy and JDBC, you should start it and make sure that it is running properly. The following section gives you a walkthrough to successfully start and test your database gateway with ShardingSphere.

Step 1 – Read/write splitting

In this case, we are going to create a read/write splitting strategy by inserting data into two replica databases, and then *watermark* the data in each database so that you can know which replica database the data comes from.

We will do so by inserting SQL first into each replica database one by one, and then query the system to confirm the routing node. For this procedure, you can base your procedure on the following code:

1. Here is the SQL script:

```SQL
\c demo_primary_ds;
CREATE TABLE t_order (
    order_id INT PRIMARY KEY NOT NULL,
    user_id INT NOT NULL,
```

```
    status CHAR(10) NOT NULL
);
```

2. First, insert the data into the primary database as follows:

```SQL
INSERT INTO t_order (order_id, user_id, status)
VALUES (1, 10001, 'write'),
       (2, 10002, 'write'),
       (3, 10003, 'write'),
       (4, 10004, 'write'),
       (5, 10005, 'write'),
       (6, 10006, 'write'),
       (7, 10007, 'write'),
       (8, 10008, 'write'),
       (9, 10009, 'write'),
       (10, 10010, 'write');
```

3. Once done, you can create a replica database called `demo_replica_ds_0`:

```
\c demo_replica_ds_0;
CREATE TABLE t_order (
    order_id INT PRIMARY KEY NOT NULL,
    user_id INT NOT NULL,
    status CHAR(10) NOT NULL
);
```

4. Now that a replica database is created, you can insert the data in the demonstration replica database as well:

```
INSERT INTO t_order (order_id, user_id, status)
VALUES (1, 20001, 'read_0'),
       (2, 20002, 'read_0'),
       (3, 20003, 'read_0'),
       (4, 20004, 'read_0'),
       (5, 20005, 'read_0'),
       (6, 20006, 'read_0'),
       (7, 20007, 'read_0'),
```

```
        (8, 20008, 'read_0'),
        (9, 20009, 'read_0'),
        (10, 20010, 'read_0');
```

5. Next, we create a second replica database called `Demo_replica_ds_1` as follows:

```
\c demo_replica_ds_1;
CREATE TABLE t_order (
    order_id INT PRIMARY KEY NOT NULL,
    user_id INT NOT NULL,
    status CHAR(10) NOT NULL
);
```

6. We again insert the data into the newly created replica database:

```
INSERT INTO t_order (order_id, user_id, status)
VALUES (1, 30001, 'read_1'),
        (2, 30002, 'read_1'),
        (3, 30003, 'read_1'),
        (4, 30004, 'read_1'),
        (5, 30005, 'read_1'),
        (6, 30006, 'read_1'),
        (7, 30007, 'read_1'),
        (8, 30008, 'read_1'),
        (9, 30009, 'read_1'),
        (10, 30010, 'read_1');
```

Now, initiate `ShardingSphere-proxy`, log in, and execute the following SQL statements:

```SQL
readwrite_splitting_db=> SELECT order_id, user_id, status FROM
t_order;
readwrite_splitting_db=> SELECT order_id, user_id, status FROM
t_order;
```

Check the ShardingSphere log to confirm whether the routing node is `replica_ds_0` or `replica_ds_1`.

Step 2 – Replica disable

With the completion of step 1, `ShardingSphere-Proxy` now has the read/ write splitting capabilities. We will move on to the next step, where we will use DistSQL to check the status of the primary and replica databases:

```SQL
readwrite_splitting_db=> SHOW READWRITE_SPLITTING READ
RESOURCES;
    resource    |  status
----------------+----------
  replica_ds_0  | enabled
  replica_ds_1  | enabled
(2 rows)
```

You can disable `replica_ds_1`:

```SQL
readwrite_splitting_db=> DISABLE READWRITE_SPLITTING READ
replica_ds_1;
```

Here is the status after disabling:

```SQL
readwrite_splitting_db=> SHOW READWRITE_SPLITTING READ
RESOURCES;
    resource    |  status
----------------+----------
  replica_ds_0  | enabled
  replica_ds_1  | disabled
(2 rows)
```

Now, execute SQL to verify that ShardingSphere routing is correct. Check the ShardingSphere log to confirm that the routing node is `replica_ds_0`:

```SQL
readwrite_splitting_db=> SELECT order_id, user_id, status FROM
t_order;
```

Step 3 – Circuit breaker

If you'd like to implement the circuit breaker feature in addition to the read/write splitting feature, as configured in the previous two steps, you can refer to the following steps. You should note that the circuit breaker feature can be configured independently of whether you implement read/write splitting or not.

To configure the circuit breaker, start by initiating two ShardingSphere-Proxies: ShardingSphere-Proxy-3307 and ShardingSphere-Proxy-3308. Log in to ShardingSphere-Proxy-3307 and check the current instance list:

```SQL
readwrite_splitting_db=> SHOW INSTANCE LIST;
     instance_id       |       host       | port | status   | labels
-----------------------+------------------+------+----------+--------
  192.168.2.184@3308 | 192.168.2.184 | 3308 | enabled |
  192.168.2.184@3307 | 192.168.2.184 | 3307 | enabled |
(2 rows)
```

Disable ShardingSphere-Proxy-3308 proxy instances through ShardingSphere-Proxy-3307:

```SQL
readwrite_splitting_db=> DISABLE INSTANCE 192.168.2.184@3308;
```

Check the instance list after disabling the instance:

```SQL
readwrite_splitting_db=> SHOW INSTANCE LIST;

     instance_id       |       host       | port | status   | labels
-----------------------+------------------+------+----------+--------
  192.168.2.184@3308 | 192.168.2.184 | 3308 | disabled|
  192.168.2.184@3307 | 192.168.2.184 | 3307 | enabled |
(2 rows)
```

Log in to ShardingSphere-Proxy-3308 and verify that the current instance has been disabled:

```SQL
readwrite_splitting_db=> SHOW INSTANCE LIST;
```

```
ERROR 1000 (C1000): Circuit break mode is ON.
```

You are now familiar with ShardingSphere's database gateway feature. Consider saving the examples you found in this section and coming back to them whenever you are having trouble or need to set up a database gateway in the future.

Summary

In this chapter, through carefully selected cases, you got a complete overview of some real-world ShardingSphere application scenarios. These cases have been chosen based on our experience and Apache ShardingSphere's community-rich experience in working closely with stock market-listed enterprises handling billions of rows of often highly-sensitive data.

Not only do you now possess knowledge accrued thanks to the experience of large enterprises, but you are also reaching the conclusion of this book – and becoming an Apache ShardingSphere master. We suggest trying to implement the knowledge you have learned so far in your environment. If you encounter any difficulties, we recommend connecting with our community, which means you'll probably get to know us, as well.

Communities are key for us. The community has been the key to building the Apache ShardingSphere ecosystem, and a key to getting market-leading enterprises to adopt ShardingSphere. The next chapter will give a more comprehensive understanding of our community, its history, and future direction.

Appendix and the Evolution of the Apache ShardingSphere Open Source Community

Reading this book, you're probably at least intrigued by Apache ShardingSphere and eager to gain a better understanding of this open source project. If you are browsing this appendix after completing the previous chapters, then it means that you have officially reached the end of this book and are ready to master Apache ShardingSphere and take any database to which you apply ShardingSphere to the next level.

If you haven't gone through this book in a linear fashion and have jumped to the appendix first, you are in the right place too.

To make the book informative, we provided plenty of information concerning project descriptions, configuration, tests, and use cases, for both beginners and advanced learners.

The previous 12 chapters were divided into three main parts. Except for the first four chapters, you can choose to read or revisit the following chapters independently, depending on your interests and time:

- In *Chapters 1-4*, we introduced the basics of Apache ShardingSphere.
- In *Chapters 5-7*, we discussed the common but classic applications of Apache ShardingSphere.
- In *Chapters 8-12*, we shared the advanced applications for Apache ShardingSphere.

This book covered all the basic aspects of Apache ShardingSphere as inclusively as possible, making it a great way to understand its technology. To help you further study Apache ShardingSphere, we also provided advanced applications, test cases, and use cases for your reference.

Although Apache ShardingSphere is updated very quickly, and this book is based on version 5.0.0 to enhance your understanding and use of the product – its inclusive community, plugin-oriented architecture, and ecosystem-supporting concept will never change.

This chapter will guide you through the following topics:

- How to leverage the project documentation
- How to use the example projects we offer on GitHub
- More information on Apache ShardingSphere's source code, license, and version
- An overview of our open source community and how you can join us

By the end of this chapter, you will have fully assimilated everything there is to know about Apache ShardingSphere, and you'll be able to start your open source journey.

How to leverage the documentation to find answers to your questions

This book organizes Apache ShardingSphere's concepts to help you to master all aspects from various perspectives. Let's recall the key points of each chapter.

Let's start with the basics. *Chapters 1-4* help you gain an understanding of the current industry status, its evolution, why Apache ShardingSphere was first conceived, and where it fits in the database landscape.

The second section of the book, including *Chapters 5-7*, moved you on to getting started with ShardingSphere Proxy and JDBC adaptors, configurations, their mixed deployment, and more.

The third section, from *Chapters 8-10* (once you assimilated the features, clients, and their configurations), introduced you to the concept of Database Plus and the advanced use of Apache ShardingSphere.

The final section of the book (*Chapters 11 and 12*) was all about putting theory to practice. In the chapters in this section, you were able to apply your ShardingSphere knowledge with some advanced examples. The examples reviewed in this section are all borne out of real enterprise scenarios, thanks to the feedback we collected from our enterprise users.

Now that we have reviewed the overall content of each section, helping you tie the chapters together, we should cover some auxiliary content that you will find useful. In the following section, we start by exploring the `shardingsphere-example` project.

Example project introduction

`shardingsphere-example` is an independent Maven project in the `-examples` directory of the Apache ShardingSphere project. The link to the Maven project is `https://github.com/apache/shardingsphere/tree/master/examples`.

The `shardingsphere-example` project contains multiple modules providing use cases for `ShardingSphere-JDBC`, `ShardingSphere-Proxy`, and `ShardingSphere-Parser`, for instance:

- `shardingsphere-jdbc-example`: This project gives you usages and configuration examples of functions such as data sharding, read/write splitting, encryption, shadow databases, distributed governance, and distributed transactions, as well as access methods such as Java API, YAML, Spring Boot, and Spring Namespace.

    You can refer to the directory structure as follows:

```
├── mixed-feature-example
│   └── sharding-readwrite-splitting-example
├── single-feature-example
│   ├── cluster-mode-example
│   ├── encrypt-example
│   ├── extension-example
│   ├── readwrite-splitting-example
│   ├── shadow-example
│   ├── sharding-example
│   ├── target
│   └── transaction-example
```

- `shardingsphere-proxy-example`: ShardingSphere-Proxy shows you how to configure `server.yaml` and schema rules in ShardingSphere-Proxy. Since the configuration formats of data sources and rules are consistent with ShardingSphere-JDBC, they are not repeated in the project. Instead, we demonstrate the unique Proxy functions such as `DistSQL` and `SQL Hint`.

  The directory structure you'll be presented with is as follows:

  ```
  ├── shardingsphere-proxy-boot-mybatis-example
  ├── shardingsphere-proxy-distsql-example
  └── shardingsphere-proxy-hint-example
  ```

- `other-example`: The `other-example` project contains use cases of `SQLParserEngine`, of which you have learned the concepts in the previous chapters of this book. It is the SQL parsing engine, which is customized by Apache ShardingSphere and fundamental to features of ShardingSphere-JDBC and ShardingSphere-Proxy. With `SQLParserEngine`, a SQL statement can be parsed into an abstract syntax tree, which allows programmers to develop various advanced features.

  The directory structure is as follows:

  ```
  └── shardingsphere-parser-example
  ```

Now that you have an overview of the `shardingsphere-example` project, let's explore how you can make the best use of it.

How to use the example project section

In this section, we give several typical examples to show you how to configure and run `shardingsphere-example`.

Though there are many modules in the `shardingsphere-example` project, we can only introduce some of the most popular application scenarios of ShardingSphere-JDBC.

In the following section, you will find a preparation phase that you can commonly refer to, followed by the steps that are specific to the example you are interested in. You will find that we have prepared and divided the steps into subsections according to the cases, including `sharding-springboot-mybatis` and `readwrite-splitting-raw-jdbc`.

Preparation

Maven is the project's build tool for `shardingsphere-example`. So, please set up Maven first before proceeding to the following step:

1. Prepare Apache ShardingSphere. If you have not configured Apache ShardingSphere yet, please clone and compile it first. You can use the following code to perform this step:

    ```Bash
    ## Clone Apache ShardingSphere
    git clone https://github.com/apache/shardingsphere.git
    ## Compile
    cd shardingsphere
    mvn clean install -Prelease
    ```

    > **Note**
    >
    > If you see the **Filename too long** warning when cloning a project in a Windows environment, please refer to this solution: `https://shardingsphere.apache.org/document/5.0.0/en/reference/faq/#22-other-in-windows-environmentwhen-cloning-shardingsphere-source-code-through-git-why-prompt-filename-too-long-and-how-to-solve-it.`

2. Import the `shardingsphere-example` project to your IDE.

3. Prepare a manageable database environment, such as local MySQL examples.

4. If you need to test read/write splitting, please make sure that your primary-replica synchronization is running properly.

5. Execute the `init` database script: `examples/src/resources/manual_schema.sql`.

This concludes the preparation phase; let's now move on to the actual examples.

Scenarios and examples

In this section, we will go over various examples and scenarios. The first example includes data sharding, while the second refers to the read/write splitting feature.

sharding-spring-boot-mybatis-example

This example shows you the application of ShardingSphere-JDBC in combination with SpringBoot and MyBatis to perform sharding. The goal is to shard one table's data into four tables evenly and store them in two different databases.

The path is as follows:

```
examples/shardingsphere-jdbc-example/single-feature-example/
sharding-example/sharding-spring-boot-mybatis-example
```

Let's get prepared for our goal:

1. Configure `application.properties`.
2. Set `spring.profiles.active` as `sharding-databases-tables`.
3. Configure `application-sharding-databases-tables`.
4. Change `jdbc-url` with your database location and set up your user ID, password, and so on.
5. Set the attribute of `spring.shardingsphere.props.sql-show` as `true`.

See more details in *Chapter 5, Exploring ShardingSphere Adaptors*.

Now, let's run the `ShardingSpringBootMybatisExample.java` startup.

Now, you can observe the routing of all SQL statements in Logic SQL and Actual SQL logs and understand how sharding works.

readwrite-splitting-raw-jdbc-example

This example shows how you can use YAML and configure the feature read/write splitting of ShardingSphere-JDBC. The goal is to separate one writing database and two reading databases.

The path is as follows:

```
examples/shardingsphere-jdbc-example/single-feature-example/
readwrite-splitting-example/
readwrite-splitting-raw-jdbc-example
```

Let's get prepared for our goal:

1. Configure `resources/META-INF/readwrite-splitting.yaml`.

2. Change `jdbc-url` with your database location and set up your user ID, password, and so on.

3. Set `props.sql-show` as `true`.

See more details in *Chapter 5, Exploring ShardingSphere Adaptors*.

Now, run the `ReadwriteSplittingRawYamlConfigurationExample.java` startup.

Now, you can observe the routing of all SQL expressions in Logic SQL and Actual SQL logs and understand how read and write splitting works.

> **Note**
> When primary-replica database synchronization fails, there will be query errors.

The `shardingsphere-example` project includes examples such as the ones introduced in this section, but also many more. It is a community-driven effort, and you will find other useful examples on the Apache ShardingSphere GitHub repository if you'd like to browse additional examples.

Source code, license, and version

Let's start by reviewing the source code and some common terminology. You can leverage the following section as a glossary. We have classified the terms according to their respective modules.

In the following two subsections, we will proceed to review the current version and the Apache ShardingSphere license.

shardingsphere-kernel

The `shardingsphere-kernel` module consists of `authority`, `single-table`, and `transaction`. When using ShardingSphere, you will become pretty accustomed to these terms and components. The following is a refresher of the definitions that we covered in the previous chapters of this book:

The `authority` module is used to allocate and manage user authority. It includes submodules of `api` and `core`:

- `api` provides a universal interface for authority configuration items and relevant authorities.
- The `core` module provides authority verification, authority algorithm, and authority rules among other implementation items.

The `single-table` module is responsible for the use of single-table databases, and it includes `api` and `core`:

- The `api` module provides default configuration items of a single table.
- The `core` module can implement functions such as single-table rule loading, metadata loading, and single-table routing.

The `transaction` module is used for managing transactions. It includes `api`, `core`, and `type`:

- The `api` module provides the default configuration item of the transaction.
- The `core` module can implement event rules, transaction manager, and transaction functions.
- The `type` module includes `base` and `xa`:
  - The `base` module includes `seata-at`: The `seata-at` module provides the transaction manager of the `seata-at` mode and enables the implementation of relevant functions of the `seata-at` mode.
  - The `xa` module includes `core`, `provider`, and `spi`:
    - The `core` module: This module enables the implementation and abstraction of `xa` transaction functions.
    - The `provider` module includes `atomikos`, `bitronix`, and `narayana`: The `atomikos` module can implement `atomikos` transactions. The `bitronix` module can implement `bitronix` transactions. The `narayana` module enables relevant implementation of the `narayana` transaction.
    - The `spi` module provides an SPI loading interface of `xa` transactions, enabling users to implement `xa` transactions themselves.

shardingsphere-infra

The following list is a refresher of the shardingsphere-infra module, which contains binder, common, context, datetime, executor, merge, optimize, parser, rewrite, and route:

- binder is mainly used to set context information in SQL statements.

- common contains the abstraction and relevant port definition of general functions of ShardingSphere. More specifically, it has functions such as algorithm definition, mode definition, rule definition, database configuration, metadata information, event delivery, abnormality, and YAML configuration conversion.

- context mainly provides functions such as kernel execution process management and metadata refresh.

- datetime provides database time and system time.

- executor provides relevant functions to execute SQL, such as SQL checker and execution engine.

- merge provides relevant functions for merging SQL result sets by way of stream result set merger and memory result set merger.

- optimize provides relevant functions of the query optimizer. It is mainly used to support and optimize complex SQL.

- parser provides the function of parsing SQL statements and can improve parsing performance by its caching capability.

- rewrite provides the function of rewriting SQL statements and provides universal, basic components for rewriting sharding, encryption, and decryption functions.

- route provides the function of SQL routing, constituting a universal, basic component of SQL routing.

shardingsphere-jdbc

The shardingsphere-jdbc module provides a ShardingSphere API that has reached JDBC standard, and provides an easy project access way that can use the Spring framework:

- shardingsphere-jdbc-core provides a ShardingSphere API that has reached JDBC standard.

- shardingsphere-jdbc-spring provides Spring Namespace and Spring Boot Starter.

- `shardingsphere-jdbc-transaction-spring` implements the **aspect-oriented programming** (**AOP**) processing logic of ShardingSphere transactional annotation.

- `shardingsphere-jdbc-core-spring` provides Spring Namespace processing logic and Spring Boot Starter automatic configuration logic.

- `shardingsphere-jdbc-spring-infra` provides the tools needed for Spring access.

shardingsphere-db-protocol

The `shardingsphere-db-protocol` module defines message formats of various database protocols:

- `shardingsphere-db-protocol-core` defines constants and interfaces related to database protocols.

- `shardingsphere-db-protocol-mysql` defines the `MySQL` protocol message format.

- `shardingsphere-db-protocol-postgresql` defines the `PostgreSQL` protocol message format.

- `shardingsphere-db-protocol-opengauss` defines the `openGauss` protocol message format.

shardingsphere-proxy

The `shardingsphere-proxy` module contains protocol implementation of MySQL, PostgreSQL, and openGauss databases, which are supported by ShardingSphere-Proxy, and the logic of real database interaction:

- `shardingsphere-proxy-backend` provides the logic of interactions between ShardingSphere-Proxy and real databases.

- `shardingsphere-proxy-bootstrap` is responsible for configuration loading in launching ShardingSphere-Proxy.

- `shardingsphere-proxy-frontend` implements the interaction between database protocols and is responsible for the interaction with the client ends.

- `shardingsphere-proxy-frontend-spi` defines an SPI related to protocol implementation.

- `shardingsphere-proxy-frontend-core` is responsible for the universal logic of interaction with client ends.

- `shardingsphere-proxy-frontend-mysql` implements the logic of the interaction with the MySQL protocol.

- `shardingsphere-proxy-frontend-postgresql` implements the logic of the interaction with the PostgreSQL protocol.

- `shardingsphere-proxy-frontend-opengauss` implements the logic of the interaction with the openGauss protocol.

shardingsphere-mode

The `shardingsphere-mode` module provides support for ShardingSphere's running mode. The `running` mode is a new concept introduced in Version 5.0.0, providing you with options including the `memory` mode, the `standalone` mode, and the `cluster` mode to meet your needs in various application scenarios.

The structure of the module is as follows:

```
├──shardingsphere-mode-
core                                                    # Table
Structure Configuration
├──shardingsphere-mode-type
├       ├──shardingsphere-cluster-mode
├       ├       ├──shardingsphere-cluster-mode-core
├       ├       ├──shardingsphere-cluster-mode-repository
├       ├       ├       ├──shardingsphere-cluster-mode-repository-
api
├       ├       ├       ├──shardingsphere-cluster-mode-repository-
provider
├       ├       ├       ├       ├──shardingsphere-cluster-mode-
repository-etcd
├       ├       ├       ├       ├──shardingsphere-cluster-mode-
repository-zookeeper-curator
├       ├──shardingsphere-memory-mode
├       ├       ├──shardingsphere-memory-mode-core
├       ├──shardingsphere-standalone-mode
├       ├       ├──shardingsphere-standalone-mode-core
├       ├       ├──shardingsphere-standalone-mode-repository
├       ├       ├       ├──shardingsphere-standalone-mode-
repository-api
```

```
├    ├    ├         ├──shardingsphere-standalone-mode-
repository-provider
├    ├    ├    ├         ├──shardingsphere-standalone-mode-
repository-file
```

The `shardingsphere-mode-core` module provides core and universal functions for different running modes, such as metadata persistence service and managing the context of metadata.

Three running modes are implemented under `shardingsphere-mode-type` module, and they are as follows:

```Java
shardingsphere-cluster-mode        cluster mode
shardingsphere-memory-mode         memory mode
shardingsphere-standalone-mode     standalone mode
```

Since the `memory` mode does not require metadata persistence, its implementation is the simplest. It only includes the `shardingsphere-memory-mode-core` module and can create a context of metadata under the `memory` mode.

The `cluster` and `stand-alone` modes have similar module components, which are `core` and `repository` modules:

- `core` has the same positioning as the `memory` module and it is responsible for creating metadata context. Besides providing metadata persistence, the `cluster` mode also needs to support distributed governance function. Therefore, the `core` module also contains processing logic for cluster synchronization events, ensuring the dynamic online upgrading of metadata.

- The `repository` module defines the persistence plan under the current running mode, the `api` module defines necessary SPI access, and the `provider` module defines a detailed persistence solution. ShardingSphere has built-in persistence solutions for different modes, allowing you to use them by adding configuration. The `cluster` mode provides default implementation based on etcd and ZooKeeper, and the `stand-alone` mode provides implementation based on local documents.

shardingsphere-features

Let us look at the modules that are classified according to their features in the ShardingSphere ecosystem. You will find database discovery, encryption, read/write splitting, shadow, sharding, agent, DistSQL, SPI, and test and distribution features.

db-discovery

The `shardingsphere-db-discovery` module contains `api`, `core`, `distsql`, and `spring`:

- `api` provides rule configuration and SPI access for database discovery.
- `core` provides the implementation of the database discovery rule, heartbeat detection at a regular time, validator, and routing rules.
- The `distsql` module includes `handler`, `parser`, and `statement`:
  - `handler` is responsible for processing relevant `distsql` module of database discovery, including showing and modifying database discovery.
  - `parser` is responsible for parsing related `distsql` module of database discovery.
  - `statement` contains items of the `distsql` parsing results.
- The `spring` module contains `spring-boot-starter` and `spring-namespace`:
  - `spring-boot-starter` provides access to the relevant configuration of database discovery based on Spring Boot.
  - `spring-namespace` provides access to the relevant configuration of database discovery based on XML.

encrypt

The `shardingsphere-encrypt` module contains `api`, `core`, `distsql`, and `spring`:

- `api` includes rule configuration and SPI access to the encryption algorithm and decryption algorithm.
- `core` contains the implementation of the **AES**, **MD5**, and **RC4** algorithms, validator, encryption rule, encrypted table metadata loading, rewriting encryption and decryption, and merging encryption and decryption results.
- `distsql` includes `handler`, `parser`, and `statement`:
  - `handler` is responsible for processing `distsql` module related to `encrypt`, including showing and modifying encryption rules.
  - `parser` is responsible for parsing `distsql` related to `encrypt`.
  - `statement` contains items of the `distsql` parsing results.

- The `spring` module contains `spring-boot-starter` and `spring-namespace`:

  - `spring-boot-starter` provides access to configurations related to `encrypt` based on `spring-boot-starter`.

  - `spring-namespace` provides access to configurations related to `encrypt` based on `spring-namespace`.

read/write-splitting

The `shardingsphere-readwrite-splitting` module contains `api`, `core`, `distsql`, and `spring`:

- `api` provides read/write splitting rule configuration and SPI access to the load balancing algorithm.

- `core` provides the implementation of read/write splitting rule and the load balancing algorithm, including the implementation of validator and routing rules.

- `distsql` contains `handler`, `parser`, and `statement`:

  - `handler` is responsible for processing `distsql` related to read/write splitting, including showing and modifying read/write splitting rules.

  - `parser` is responsible for parsing `distsql` related to read/write splitting.

  - `statement` contains items of the `distsql` parsing results.

- `spring` contains `spring-boot-starter` and `spring-namespace`:

  - `spring-boot-starter` provides access to relevant configurations of read/write splitting based on `spring-boot-starter`.

  - `spring-namespace` provides access to relevant configurations of read/write splitting based on `spring-namespace`.

shadow

The `shardingsphere-shadow` module contains `api`, `core`, `distsql`, and `spring`:

- `api` provides shadow database rule configuration and SPI access to the shadow algorithm.

- `core` provides the shadow database rule and the implementation of the shadow algorithm, including the implementation of routing rules and validator.

- `distsql` includes `handler`, `parser`, and `statement`:

  - `handler` is responsible for processing `distsql` related to the shadow database, including showing and modifying shadow rules.

  - `parser` is responsible for parsing `distsql` related to the shadow database.

  - `statement` contains items of the `distsql` parsing results.

- `spring` contains `spring-boot-starter` and `spring-namespace`:

  - `spring-boot-starter` provides access to relevant configurations of the shadow database based on `spring-boot-starter`.

  - `spring-namespace` provides access to relevant configurations of the shadow database based on `spring-namespace`.

sharding

The `shardingSphere-sharding` module contains `api`, `core`, `distsql`, and `spring`:

- `api` contains relevant configuration items of sharding rules, strategy, and type, as well as generates SPI access for the sharding algorithm and primary key.

- `core` mainly provides the implementation of the algorithm, validator, and rules. It also provides the implementation of loading, routing, rewriting, and merging metadata.

- `distsql` contains `handler`, `parser`, and `statement`:

  - `handler` is responsible for processing `distsql` related to sharding, including showing and modifying sharding rules and algorithms.

  - `parser` is responsible for parsing `distsql` related to sharding.

  - `statement` contains items of the `distsql` parsing results.

- `spring` contains `spring-boot-starter` and `spring-namespace`:

  - `spring-boot-starter` provides access to relevant sharding configurations based on `spring-boot-starter`.

  - `spring-namespace` provides access to relevant sharding configurations based on `spring-namespace`.

shardingsphere-agent

shardingSphere-agent includes api, core, plugins, bootstrap, and distribution:

- api provides kernel interface and abstraction items used by the agent, as well as universal configuration items.

- core module encapsulates the general capabilities of bytecode injection and provides a loading and configuration process for agent plugins, which is the basis for the work of the agent module. It also provides loading for plugin configuration and JAR add-on files.

- plugins includes logging, metrics, and tracing. It implements data gathering of observable indicators based on core modules:

  - logging provides a simple function to intercept and input log.

  - metrics implements the function of gathering often-used metrics with observability based on prometheus.

  - tracing implements the function of link data gathering based on **Zipkin**, **Jaeger**, **OpenTelemetry**, and **SkyWalking OpenTracing**. The tracing-test module is a framework providing a unit testing function for the tracing module.

- bootstrap mainly provides integration for Java agent and agent access.

- distribution provides the function of packaging. shardingsphere-agent packaging files will be output to the target directory of this module.

shardingsphere-sql-parser

The shardingsphere-sql-parser module contains dialect, engine, spi, and statement:

- dialect provides parsing for various database dialects such as MySQL, openGauss, Oracle, PostgreSQL, SQL Server, and other database dialects that use SQL92 syntax.

- engine provides an SQL parsing engine, which can parse SQL, and contains a cached syntax tree that can be used to improve performance.

- spi provides abstraction access for parser and visitor, as well as SPI access for the implementation of parsing other database dialects.

- statement provides various segments for storing parsing results, and provides some universal solutions for receiving parsed segments.

shardingsphere-distsql

In Apache ShardingSphere, the `shardingsphere-distsql` module provides basic functions of DistSQL, including a DistSQL syntax definition document, parsing capability, and parsed statement objects corresponding to the syntax. It mainly includes the following modules:

- `shardingsphere-distsql-parser`: This module provides the DistSQL syntax definition and the capability of parsing DistSQL. `shardingsphere-distsql-parser` provides unified parsing access for any DistSQL statement input. Different types of DistSQL will be routed to different parsers, generating statement objects.

- `shardingsphere-distsql-statement`: This module provides objects corresponding to DistSQL type, and it will process DistSQL according to statement objects.

shardingsphere-spi

In Apache ShardingSphere, the `shardingsphere-spi` module provides all SPI access and the capability of loading an SPI, representing the foundation of Apache ShardingSphere's pluggable capability. The module does not have classifications and only divides an SPI in Apache ShardingSphere into `optional`, `ordered`, `required`, `singleton`, and `typed`, according to their packages. The detailed structure is as follows:

```
├── exception
├── optional
├── ordered
├── required
├── singleton
└── typed
```

shardingsphere-test

The `shardingsphere-test` module provides integrated testing support for the function points and the application scenarios of multiple functions of ShardingSphere:

```
shardingsphere-test
    ├──shardingsphere-integration-agent-test
    ├      ├──shardingsphere-integration-agent-test-plugins
    ├      ├        ├──shardingsphere-integration-agent-test-metrics
```

```
├    ├        ├──shardingsphere-integration-agent-test-common
├    ├        ├──shardingsphere-integration-agent-test-zipkin
├    ├        ├──shardingsphere-integration-agent-test-jaeger
├    ├        ├──shardingsphere-integration-agent-test-
opentelemetry
├──shardingsphere-integration-driver-test
├──shardingsphere-integration-scaling-test
├        ├──shardingsphere-integration-scaling-test-mysql
├──shardingsphere-integration-test
├        ├──shardingsphere-integration-test-fixture
├        ├──shardingsphere-integration-test-suite
├──shardingsphere-optimize-test
├──shardingsphere-parser-test
├──shardingsphere-pipeline-test
├──shardingsphere-rewrite-test
├──shardingsphere-test-common
```

Let us look at the applications of these functions:

- `shardingsphere-integration-agent-test`: Provides integrated testing for `shardingsphere-agent`

- `shardingsphere-integration-driver-test`: Provides integrated testing for ShardingSphere's support for different databases and connection pools

- `shardingsphere-integration-scaling-test`: Provides integrated testing for `shardingsphere-data-pipeline`

- `shardingsphere-integration-test`: Provides integrated testing for single function and function combinations of `shardingsphere-features`

- `shardingsphere-optimize-test`: Provides integrated testing for `shardingsphere-infra-federation-optimizer`

- `shardingsphere-parser-test`: Provides integrated testing for `shardingsphere-sql-parser`

- `shardingsphere-pipeline-test`: Provides integrated testing for `shardingsphere-data-pipeline`

- `shardingsphere-rewrite-test`: Provides integrated testing for `shardingsphere-infra-rewrite`

- `shardingsphere-test-common`: Contains `MockedDataSource`, and is the common toolkit for all tests that need a mocked data source

shardingsphere-distribution

The `shardingsphere-distribution` module is used to create source code packages and ShardingSphere-JDBC and ShardingSphere-Proxy binary distribution packages:

- `shardingsphere-jdbc-distribution` will create a `tar.gz` package out of the JAR of relevant modules of ShardingSphere-JDBC (except for reliance on a third party).

- `shardingsphere-proxy-distribution` will create a `tar.gz` package out of the JAR of relevant modules of ShardingSphere-Proxy and reliance on a third party.

- `shardingsphere-src-distribution` will create a `.zip` package out of ShardingSphere project source codes (except for documents, examples, and CI configuration).

This concludes your overview of ShardingSphere's source code modules. The next two sections will give you a brief overview of ShardingSphere's license and introduction.

License introduction

ShardingSphere uses the Apache Software Foundation's license. The Foundation offers multiple licenses to distribute both software and documentation, while our community uses what is known as the **Apache License 2.0**.

This license was approved by the Foundation in 2004 and is part of what is known as *permissive* licenses, as opposed to *copyleft* licenses.

Copyleft licenses are so called because they require any code that is built as derivative work of the copyleft licensed software to be released under the same license as the original software.

On the other hand, permissive licenses, such as the Apache License 2.0, do not have any such requirement. This means that as a user using software such as ShardingSphere, you can do nearly anything with the code. Examples include sublicensing the code, including it in a commercial product you're building, distributing any copies or modifications of the code, or simply altering the code.

For a complete copy of the Apache License 2.0, you can consult the official Foundation's website at the following link: `https://www.apache.org/licenses/LICENSE-2.0`.

Version introduction

This book is based on Apache ShardingSphere version 5.0.0. The road to arriving at this version has been a long one, full of efforts, volunteer dedication, and sheer passion for open source software.

By the time this book will be released, our community will probably have released version 5.1.1. This is because, since the end of 2021, our community has opted to increase the frequency of releases.

If you consult our official website, you will see that the time period between releases has significantly been shortened. We have collectively opted for the path of frequent iteration in order to better absorb community feedback, and better share the workload across the community.

Version 5.0.0 is an especially important one, as it marked Apache ShardingSphere's evolution from middleware to a fully-fledged ecosystem capable of satisfying any database requirement and enhancing it with new features. We strive to maintain complete and historically-accurate documentation, and you will be able to find documentation for all of ShardingSphere's legacy versions on our official website.

Considering that you may or may not visit our website and GitHub repository in the future, we believe that you might want to take the leap and become an open source contributor. The next section will give you an overview of our open source community and our history, and introduce you to open source contributions in general.

Open source community

As a community, we take pride in our history, as relatively short as it may be. Our open source project may indeed be relatively young if compared to other Apache projects, but we believe that the successes we have achieved in such a short time stand as a testament to our commitment and dedication to open source.

You can trace our roots to 2015 when the JDBC project was first started. Since then, we introduced the project to the Apache Software Foundation's incubator in 2018 and quickly graduated as an Apache Top-Level Project in 2020.

The following list provides you with an overview of some important dates for our community:

- **October 2015**: Sharding-JDBC project initiation.
- **January 18, 2016**: Sharding-JDBC officially open sourced.
- **May 10, 2015**: Renamed as ShardingSphere.

- **November 10, 2018**: ShardingSphere project enters the Apache incubator.
- **December 21, 2018**: The first Apache ShardingSphere (incubating) **process and project management committee (PPMC)** meeting.
- **April 21, 2019**: Apache ShardingSphere version 1.0.
- **April 16, 2020**: ShardingSphere crowned Apache Top-Level Project.
- **June 2020**: ElasticJob becomes a subproject.
- **June 19, 2021**: ShardingSphere 5.0.0-beta version.
- **July 6, 2021**: ElasticJob 3.0 version.
- **September 17, 2021**: Four awards are given by CAICT at OSCAR Open Source Industry Conference.
- **October 12, 2021**: ElasticJob 3.0.1 version.
- **November 10, 2021**: ShardingSphere 5.0.0 GA version.
- **November 13, 2021**: China ShardingSphere open source community launched.
- **December 11, 2021**: The first Apache ShardingSphere dev meetup.
- **February 16, 2022**: Apache ShardingSphere 5.1.0 version.

As you can see, the development has been fairly fast. For a more extensive roadmap, you can consult our official website. The next list provides you with some of our community milestones:

- **2016**: *1.x Version Enhanced Database Driver* | 1,455 Stars | 7 Contributors
- **2017**: *2.x Version Distributed Governance* | 3,134 Stars | 15 Contributors
- **2018**: *3.x Version Database Proxy & ShardingSphere Project Enters the Apache Incubator* | 5,992 Stars | 44 Contributors
- **2020**: *4.x Version Became One of the Top-Level Apache Projects* | 12,964 Stars | 208 Contributors
- **2021**: *5.x Version To Pluggable and The Database Plus Concept is Born* | 14,612 Stars | 302 Contributors
- **2023**: *6.x Version To Developer Simplified SPI Makes it Easier for Developers to Contribute and Use ShardingSphere*
- **2025**: *7.x Version To Ecosystem Provide More Functions and Support More Databases Build Upper-Level Ecosystem and Standard for Heterogeneous Database*

Our community is focused on expanding and continuously improving. For these reasons, we keep an up-to-date milestones list on our GitHub page if you want the latest updates.

Open source contribution

Contributing to open source is certainly an opportunity you should consider. The advantages it entails are many and include learning or improving your skills, connecting with great people from all over the world, and raising your professional profile. In fact, it was open source that brought the three authors of this book together and created the professional opportunities we are currently working on.

You can contribute by submitting your code to the ShardingSphere repository on GitHub, but there are other contribution avenues as well. You can contribute with documentation writing or translation, website building or design, community building efforts, social media management, and more. We believe that the focus should not be placed on what someone contributes – for us, first and foremost, what matters is the passion for open source. Being interested in our project and dedicating some of your precious time to making our community better, and ultimately advancing the project, makes you a contributor. It can range from asking questions, to providing new features as code patches. Contributions can be made in many ways.

The first step is to get involved. Dive into the Apache ShardingSphere concept, say hi to the community, and then get to action by either working on the documentation or contributing code supporting other users.

You might just be surprised how welcoming and open folks are in the open source world or the ShardingSphere community. You will soon find that the more visible and engaged you are in a project, the more fun you will have and the more access you will have to valuable advice and feedback.

The community's structure is very flat, although you will see that we have a **project management committee** (**PMC**), which can be accessed by nomination and voting once you get selected to rank up from simple contributor to committer.

As a *community*, we follow the Apache community's process for accepting a new committer. The contributors that are most active get invited by the PMC and other committers to become a committer. The perks associated with being a committer include the ability to write access to the project repository and resources, not just read source code files.

A *committer* has the privilege to write and commit code and create pull requests to the default branch of the repository and, in most cases, reviewers are also committers, as you need to understand the code to review a pull request. In other words, a committer is authorized to merge user contributions into the project.

Website and documentation

In this section, we will provide you with a series of useful links and resources to consult or join our community and more. We are constantly expanding the resources available online to the community to facilitate the understanding of what Apache ShardingSphere is and are constantly expanding the ways we engage as a community. For example, most recently, we are starting to gather both virtually and in person as a community for developer meetups.

Websites

If you search for Apache ShardingSphere on the most popular search engines, you will find that we are expanding across multiple platforms and websites – after all, this book is the perfect example of our community-building efforts.

You will also find some websites that are in various languages, and if interested, feel free to start your own in your preferred language. Nevertheless, for the *official* channels operated directly by our community, you can refer to the following list:

- Apache ShardingSphere official website: `https://shardingsphere.apache.org`

- GitHub official repository: `https://github.com/apache/shardingsphere`

- ShardingSphere's official blog: `https://shardingsphere.apache.org/blog/en/material/`

- ShardingSphere's official medium blog: `https://shardingsphere.medium.com`

Channels

If you'd like to connect with us, and we sincerely hope that you do, you can find us and other community members on these channels. We keep in touch on social media and have recently started an official **Slack Community** channel for quicker discussions or troubleshooting:

- Twitter: `@ShardingSphere`

- LinkedIn: `https://www.linkedin.com/showcase/apache-shardingsphere`

- Slack community: `https://apacheshardingsphere.slack.com/join/shared_invite/zt-sbdde7ie-SjDqo9~I4rYcR18bq0SYTg#/shared-invite/email`

Concluding note

We hope that this appendix will prove useful to you in the future to support you in operating Apache ShardingSphere, and ultimately, that this open source project will allow you to get the best out of your databases.

If what you are looking for cannot be found in this appendix, you can drop a line to any of us or reach out to our community.

Index

`Packt.com`

Subscribe to our online digital library for full access to over 7,000 books and videos, as well as industry leading tools to help you plan your personal development and advance your career. For more information, please visit our website.

Why subscribe?

- Spend less time learning and more time coding with practical eBooks and Videos from over 4,000 industry professionals

- Improve your learning with Skill Plans built especially for you

- Get a free eBook or video every month

- Fully searchable for easy access to vital information

- Copy and paste, print, and bookmark content

Did you know that Packt offers eBook versions of every book published, with PDF and ePub files available? You can upgrade to the eBook version at `packt.com` and as a print book customer, you are entitled to a discount on the eBook copy. Get in touch with us at `customercare@packtpub.com` for more details.

At `www.packt.com`, you can also read a collection of free technical articles, sign up for a range of free newsletters, and receive exclusive discounts and offers on Packt books and eBooks.

Other Books You May Enjoy

If you enjoyed this book, you may be interested in these other books by Packt:

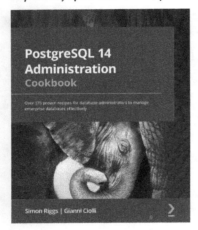

PostgreSQL 14 Administration Cookbook

Simon Riggs, Gianni Ciolli

ISBN: 9781803248974

- Plan, manage, and maintain PostgreSQL databases in production
- Work with the newly introduced features of PostgreSQL 14
- Use pgAdmin or OmniDB to perform database administrator (DBA) tasks
- Use psql to write accurate and repeatable scripts
- Understand how to tackle real-world data issues with the help of examples
- Select and implement robust backup and recovery techniques in PostgreSQL 14
- Deploy best practices for planning and designing live databases

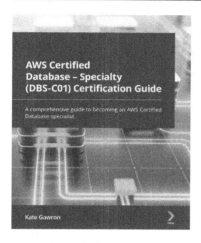

AWS Certified Database - Specialty (DBS-C01) Certification Guide

Kate Gawron

ISBN: 9781803243108

- Become familiar with the AWS Certified Database – Specialty exam format
- Explore AWS database services and key terminology
- Work with the AWS console and command line used for managing the databases
- Test and refine performance metrics to make key decisions and reduce cost
- Understand how to handle security risks and make decisions about database infrastructure and deployment
- Enhance your understanding of the topics you've learned using real-world hands-on examples
- Identify and resolve common RDS, Aurora, and DynamoDB issues

Packt is searching for authors like you

If you're interested in becoming an author for Packt, please visit `authors.packtpub.com` and apply today. We have worked with thousands of developers and tech professionals, just like you, to help them share their insight with the global tech community. You can make a general application, apply for a specific hot topic that we are recruiting an author for, or submit your own idea.

Share Your Thoughts

Now you've finished *A Definitive Guide to Apache ShardingSphere*, we'd love to hear your thoughts! Scan the QR code below to go straight to the Amazon review page for this book and share your feedback or leave a review on the site that you purchased it from.

https://packt.link/r/1-803-23942-5

Your review is important to us and the tech community and will help us make sure we're delivering excellent quality content.

www.ingramcontent.com/pod-product-compliance
Lightning Source LLC
Chambersburg PA
CBHW081457050326
40690CB00015B/2827